STEWARDSHIP ETHICS IN DEBT MANAGEMENT

Issues in Business Ethics

VOLUME 12

Stewardship Ethics in Debt Management

by

ROY MOHON

University of Sunderland,
Sunderland, U.K.

KLUWER ACADEMIC PUBLISHERS
DORDRECHT / BOSTON / LONDON

A C.I.P. Catalogue record for this book is available from the Library of Congress.

ISBN 0-7923-5747-7

Published by Kluwer Academic Publishers,
P.O. Box 17, 3300 AA Dordrecht, The Netherlands.

Sold and distributed in North, Central and South America
by Kluwer Academic Publishers,
101 Philip Drive, Norwell, MA 02061, U.S.A.

In all other countries, sold and distributed
by Kluwer Academic Publishers,
P.O. Box 322, 3300 AH Dordrecht, The Netherlands.

Printed on acid-free paper

Printed in the Netherlands.

CONTENTS

Preface *ix*

Introduction **1**

Focusing on Stewardship to Consider Debt Management 1
Stewardship and accountancy 2
Stewardship and corporate governance 3
Stewardship and environmental strategy 3
Stewardship and Christian values 4
A strategy for testing in business ethics 4
Literature resources for ethical debt management 6
The Structure of this Book 7
Notes to Introduction 9

Chapter 1
The Context of Stewardship: The Debt Problem **12**

1.1. Historical Context 12
1.1.1. Usury and scripture 12
1.1.2. The historical context 14
1.1.3. The present day 15
1.2. Moral Concerns and the Debt Problem 16
1.2.1. Economics 16
1.2.2. Business 24
1.2.3. Private individuals 27
1.3. Technical Aspects of the Morality of Debt 28
1.3.1. Economic issues 28
1.3.2. Borrowers 33
1.3.3. Lenders 37
1.3.4. Borrower-lender relationship 39
1.4. Conclusion 43
1.4.1 Factors not suitable for incorporation into an ethical model 43
1.4.2 The relevance of stewardship 44
Notes to chapter 1 47

Chapter 2
The Stewardship Concept **52**

2.1 Definitions of Stewardship in Business 52
2.1.1. Stewardship accounting 52
2.1.2. Agency theory 55
2.2. A Critical Defense of Stewardship 57
2.2.1. Stewardship distinguished from other views 57
2.2.2. Stewardship defended from current criticisms 59
2.3. Development of the Concept of Stewardship 69
2.3.1. Product stewardship 70
2.3.2. Religious stewardship 70
2.3.3. Stewarding and the environment 72
2.3.4. The stewardship concept and its core values 74
2.3.5. Key elements of stewardship for debt management 78
2.4. The Dynamics of Stewardship Decision-making 80
2.4.1. The environmental context 80
2.4.2. The market context 82
2.4.3. The business context 83
Notes to Chapter 2 88

Chapter 3
Borrowers and Borrowing **91**

3.1. Stewardship 91
3.1.1 Stewardship in existing financial management 91
3.1.2. Stewardship and ethical debt management 93
3.2. Entrustedness in Personal Debt Management 96
3.3 Matching 98
3.3.1. Defining the financial starting point 99
3.3.2. Measuring asset enhancement 100
3.3.3. Reliable information for strategic direction 100
3.3.4. Family stewardship 103
3.4. Effectiveness 106
3.5. Development 109
3.5.1. Resisting external pressures 111
3.5.2. Defining realistic objectives 113
3.6. Accountability 114
3.6.1. Accountable control 114
3.6.2. Domestic budgeting 115

3.6.3. Mastering budgeting problems 115
3.7. Fairness 117
3.8. Strategic Decisions for Contemporary Issues 117
3.8.1. Strategy selection 117
3.8.2. Stewardship of debt 120
Notes to Chapter 3 125

Chapter 4
Bankers, Business Finance and Consumer Lending 129

4.1. Business Finance 131
4.1.1. Principles: evaluation of the canons of lending 131
4.1.2. Practice: evaluation of lending procedures 133
4.2 Consumer Lending 141
4.2.1. Stewardship and individual appraisal 141
4.2.2. Stewardship and systems appraisal methods 145
4.2.3. Conclusions 149
Notes to Chapter 4 158

Chapter 5
Mortgage Finance and Money Advice 161

5.1. The Issue of Justice 161
5.2. Mortgage Finance 166
5.2.1. Mortgage finance and criteria of justice 166
5.2.2. Mortgage debt management 167
5.2.3. Debt dependency 173
5.3. Money Advice 175
5.3.1. Advice services and access 175
5.3.2. Stewardship values in money advice provision 176
5.3.3. Conclusion 184
Notes to Chapter 5 185

Chapter 6
The Stewardship Debt Management Model 188

6.1. The Concept of Stewardship 188
6.1.1. The four prerequisites 188
6.1.2. Stewardship concept values 190
6.2 The Stewardship Debt Management Model 191

6.2.1. Entrustedness 193
6.2.2. Matching (a) bankers 194
6.2.3. Effectiveness 194
6.2.4. Matching (b) borrowers 196
6.2.5. Development 198
6.2.6. Accountability 200
6.2.7. Fairness 204
6.2.8. Limitations of the model 205
6.2.9. Conclusion 207
Notes to Chapter 6 208

Bibliography **209**

Index **221**

PREFACE

As we move forward into the Third Millennium AD the perennial problem of unmanageable debt is still with us. As if to prove the point, in late November 1997, the Tokuyo City Bank in Japan closed down its business, reminding the world that default still stalks families, institutions and governments. It seems that little has been achieved in handling debt since 1216 when the *Magna Carta* limited the actions of bailiffs against debtors willing and able to make payment. Current literature about consumer credit, business finance and mortgages reveals the urgent need to tackle the ethics of borrowing and lending on some commonly understood and acceptable basis.

In this book, the stewardship concept familiar in accounting, corporate governance, environmental strategy and Christian social ethics is analyzed to provide a framework. The book demonstrates that analysis of the concept of stewardship provides a set of resource-related social values which shed light upon ethical issues in debt management and enable the construction of a decision support model to secure improvements in debt management practice. The values are applied to operational debt management as experienced by personal borrowers, small businesses and home buyers. The roles of both lenders and money advisers are considered in relation to this experience. The familiarity of the stewardship concept to a various sections of society provides a starting point for a financial services industry wide ethic for debt management. The stewardship values have been embedded together with technical finance and within the practitioner's world so that borrowers, bankers and money advisers will feel at home as the argument of the book unfolds.

The book could not have been written independently of the literary resources that have provided the raw materials for the ethical interaction contained in the following chapters and the author would like to acknowledge his debt to the various authors cited in this work and to the relevant copyright holders.

INTRODUCTION

There has been a renewed interest in stewardship as a suitable concept for pursuing business ethics. It is mentioned, for example, in Tom Cannon (1992) *Corporate Responsibility* as an important concept for the investigation of corporate governance. Peter Block in *Stewardship: Choosing Service Over Self-Interest* (1993) has applied it to the subject of human relations within organizations. This book *Stewardship Ethics in Debt Management* applies it to the operational level of personal and business activity using debt management operations for this purpose. Relationships between borrowers and lenders continue to be a subject of public concern so debt management provides a topical area in which to test the application of the stewardship concept and to consider its usefulness in securing ethical improvements. Debt management is important for individuals, businesses and sovereign states. All of these participate in the financial system as savings surplus units (lenders) or savings deficit units (borrowers). Whether arrangements are direct or through intermediaries one party can benefit at the expense of others so ethical moorings are needed to maintain continued social harmony.

Business ethics recognizes the need for appropriate frameworks to promote ethically sound decision taking at all levels. In *Business Ethics: The State of the Art*, R. Edward Freeman the editor explains:

> "The state of the art in business ethics is in creating frameworks, theoretical edifices, and decision techniques for linking these questions into coherent wholes."[1]

In connection with debt management it is clear that such frameworks and decision techniques are necessary to balance the interests of society, shareholders, lending officers and borrowers. The problem faced is how to develop an ethically sound decision support model that will meet the requirements of individuals and practising managers. In this book the stewardship concept will be used for this purpose.

Focusing on Stewardship to Consider Debt Management

Our interest is in how values associated with stewardship might enable us to construct a model to guide operational debt management. Stewardship has a long history. It was known in ancient Egypt and appears in the record of the Israelite Joseph where reference is made to *the man* (Hebrew *eesh*) over his house. It evidently persisted among the Israelites, and in King David's days there were stewards (Hebrew *sar*) over his possessions. Later it appears in Babylon in the name *Mel-tzar'* which has the article and describes the man's office. The term combines the Persian *mel* (wine) and *tzar* (head) meaning the overseer of the drink, and, by extension, of the household provisions. In first century Greek, stewardship appears in the term *oikonomos*, the (house-) steward or manager. In a large estate the latter was the landowner's deputy having considerable autonomy. Within the Christian Church the term was used to describe the work of ministers as administrators of spiritual mysteries and emphasized the faithfulness necessary. These terms show the links with:

(a) possessions or resources that belong to another,
(b) autonomy within a context of accountability to someone,
(c) deployment of resources to some beneficial end, and
(d) the importance of faithfulness in dealing with the entrusted resources.

The historical essence of stewardship is thus entrustedness, involving an accountable control of resources in the pursuit of goals; and, in its moral use, the resources and goals can be construed very broadly to include spiritual gifts and ends. With this in mind we will consider some contemporary uses of the concept which can be used to develop our understanding of stewardship.

STEWARDSHIP AND ACCOUNTANCY

Accountants distinguish between financial and management accounting. *Financial accountants* prepare financial reports showing company performance and financial position, whereas *management accountants* provide information to managers to facilitate planning, decision and control.[2] Financial accounting is often linked to stewardship. Pizzey states,

> "Financial accountants act as stewards, showing the financial effect of the actions of the owners and managers of the business."[3]

Rickwood and Thomas explain how the financial recording and control of transactions is referred to as stewardship.

> "The accountant has been traditionally regarded as the 'holder of the purse strings' and responsible for 'safeguarding' the assets of the business."[4]

Management accounting is usually distinguished as having a focus on deployment of resources. A.H. Millichamp states that

> "management accounting ... is concerned with information which is used by *management* in planning and controlling the business and in making

decisions."[5]

Such statements raise the question as to how the stewardship concept can be applied to the financial management of the firm. The answer is that management accounting provides financial information *to managers* so that they can *exercise* their stewardship, whereas financial accounting provides information *to stakeholders* to *report on* the effectiveness of managerial stewardship. It follows that both final accounts in company reports and cash and expense budgets used for managerial decision and control are relevant to stewardship.

STEWARDSHIP AND CORPORATE GOVERNANCE

Stewardship is encountered in corporate governance. Richard L. Daft explains the governance responsibility of managers as follows.

> "*The primary responsibility of top managers is to determine an organization's goals and strategy, and therein adapt the organization to a changing environment.*"[6]

The Companies Acts make clear that the management are responsible for running the company in the interests of the shareholders.[7] This management-shareholder relationship has been elucidated in terms of stewardship. Tom Cannon refers to the fact that stewardship has been placed at the centre of the policy agenda as a result of headline cases of failure. He explains:

> "Proper stewardship is an obligation placed explicitly on all those responsible for the goods or funds of others. It imposes a duty to exercise due diligence, i.e. care, rigour and attention, in the management and disposal of all those assets for which the officer is given responsibility."[8]

It is apparent from Cannon's use of the terms *obligation* and *responsibility* in this quotation that management theory implies a moral undergirding for the ideas associated with stewardship in relation to corporate governance.

STEWARDSHIP AND ENVIRONMENTAL STRATEGY

The public is interested in environmental impacts and shareholders have an increased awareness of the scale of potential liabilities. Ecology has become an issue which bankers and insurers have to address. The *Environmental Protection Act, 1990* has already produced a new caution among bankers concerning potential liabilities arising as a result of the contamination of land. Their concerns will doubtless place further pressure upon businesses to practice good environmental stewardship.[9]

STEWARDSHIP AND CHRISTIAN VALUES

Stewardship appears in the thinking of the Church as a concept defining relationship and responsibility. The steward stands in a *relationship* of entrustedness whereby there is a *responsibility* of diligence and faithfulness in the administration of resources. The extent of the user's autonomy depends upon the specifics of the particular relationship but it cannot override the fact that the possessions or resources belong to someone else and that the use of the resources for particular ends involves accountability. The examination of business practice in the light of stewardship values such as entrustedness, diligence and accountability thus sheds light upon the responsibilities involved in a variety of relationships where *some measure* of entrustedness exists.

These uses show that materials exist to analyze the stewardship concept so that it can be applied to operational debt management. The varied uses of the concept provide a basis to develop our thinking about the values implicit in stewardship. Subsequent interaction with empirical data will enable us to test the relevance and importance of the different values and thus develop a stewardship model for ethical debt management. By applying the values to practical situations it will be possible to determine how best to refine the model in order to make it more serviceable to practitioners.

A Strategy for Testing in Business Ethics

Managers are familiar with the scientific method. The early management theorists such as Frederick Taylor and Henri Fayol operated in the scientific tradition.[10] This requires appropriate strategies to test hypotheses and it is not impossible to incorporate a kind of testing in business ethics. Scientific enterprise should not be construed too narrowly. Historically, the first three academic faculties were the theological, the juridical and the medical. Taking the example of the juridical, it is readily seen that scientific study, in the sense of methodologies, relationships and the structuring of knowledge, is different in the Law Faculty to the Chemistry Department, but legal studies are no less academic or scientific because they do not employ *experimentation.*

In Kolb's experiential learning cycle, learning might start with concrete experiences that lead to observation and reflection. This can lead on to the formation of abstract concepts and the forming of an abstract rule that can be generalized and then the implications of concepts can be tested in new situations. Alternatively the learner can commence where the rule is received from others with its web of explanations and expectations and then test out.

> "In either case, whether the rule is received or generated out of the prior experience and reflection, its testing in new situations creates new experiences which enable consequent reflection, observation and ultimately new rules."[11]

Of the above alternatives the deductive approach involves the development of a conceptual and theoretical structure which is *subsequently* tested. It involves the selection of a concept relevant to the problem under investigation and the formulation of a hypothesis and the testing of the hypothesis by observations with a view to determining the extent of the validity of the hypothesis. In line with this our business ethics research can be viewed as follows:

(a) the statement of the problem for consideration - debt management;
(b) the proposal of a relevant concept - stewardship;
(c) the analysis of the concept in its relevance to the problem;
(d) the application to debt of the values derived from the concept;
(e) the statement and evaluation of the results and
(f) the formulation of a general model in the light of the testing.

The hypothesis could be stated as follows: *The analysis of the concept of stewardship provides a set of resource-related social values which shed light upon ethical issues in debt management enabling the construction of an operational model to secure improvements in practice.*

The testing will not be by experiment or survey but will use existing empirical and theoretical materials. In his literary research Gilbert employed a concurrence approach using textual sources as supporting evidence showing that the source material supported the viewpoint under consideration.[12] The more diverse the backgrounds of those whose views can be related to a particular interpretation the more its usefulness is demonstrated. Gilbert refers to a kind of validity, reliability and generalizability in such non-experimental research. The data used are the texts of others, and to be true to these in terms of validity, Gilbert simply seeks support for his claims in terms of what these sources have written. The reliability of what he is doing rests in his bringing into the testing others whose writings are considered reliable, to see what they have to say of relevance. The test of generalizability is satisfied if usefulness can been demonstrated in several general contexts. There is consequently no necessity for management scientists to feel uncomfortable with a literary approach to management research.[13] Literary resources exist to consider debt management. It is possible to consult the writings of bankers, consumer bodies, economic commentators, social science researchers and business ethics authorities to confirm the extent to which stewardship values would shed light on debt management situations.

LITERATURE RESOURCES FOR ETHICAL DEBT MANAGEMENT

The available literature presents a problem because little has been written that specifically addresses the ethics of operational debt management. The literature can be classified as:
Category (1) business ethics literature,
Category (2) debt management literature, and
Category (3) banking ethics literature.

Although there has been recent growth in the *Category (1)* literature, the books are either of a general nature, dealing with business ethics or corporate responsibility [14] or cover a more specific area, such as strategic management. [15] Neither focuses on debt management. The *Category (2)* literature deals with socio-economic issues providing a wealth of empirical data *relevant to* the ethics of debt management, but this has not been related to the ethical considerations raised by it [16]. There are two main exceptions. Firstly there are texts dealing with Third World debt [17], and, secondly, occasional research papers with some ethical emphasis.[18] The *Category (3)* literature has largely passed by the lending function. James J. Lynch, *Ethical Banking: Surviving in an age of Default* provides an example of banking ethics. Lynch proposes five ethical canons:

(a) legality,
(b) fairness,
(c) social legitimacy,
(d) justification of a course of action to stakeholders
(e) confidentiality.[19]

He considers banking codes and management development. There is material to stimulate thinking about debt but no focus upon lending. Lynch's canons prompt us not to overlook the broader issues of social and stakeholder interests in lending.

There is then a need for further work specifically focused upon the ethics of borrowing and lending operations. The debt management literature shows that there is a need for *ethical* debt management, but does not address it. The business ethics literature shows that there is a need for moral evaluation of business practice, but does not address *lending* operations. The debt management literature and the business ethics literature underline the *importance* of ethical debt management but they do not provide a *framework* for it at an operational level. In addressing this need, the absence of a body of literature specifically treating ethical debt management means that we must progress using two different kinds of literature doing justice to what *each* has to offer.

The Structure of this Book

What has been said accounts for the arrangement of the following chapters. In the first part of the book, which looks at the concept of stewardship in its context, we commence from a conceptual basis by analyzing stewardship. In the second part of the book, which deals with the application of the stewardship concept, we proceed by testing the application of the concept, addressing the question, Of what value is the stewardship concept in providing ethical insights concerning debt management? We conclude by considering the results and formulating an ethical debt management model in the light of the testing. The book as a whole is used to present the argument:

Chapter 1. This chapter defines the context of stewardship and the problems associated with debt management

Chapter 2. This chapter proposes stewardship as a useful concept to guide ethical action and analyses it in detail

Chapter 3, Chapter 4 and *Chapter 5*. These chapters test the relevance and utility of the stewardship values by applying them to a variety of cases

Chapter 6. This chapter presents the debt management model formulated in the light of the previous analysis and testing

Internalizing Values
It will be helpful at this point to mention the importance of the internalization of the stewardship values by borrowers and lending practitioners as a pre-requisite for using the ethical decision support model. Our model is very different to a statistically based model such as those used for credit scoring. The latter require the appropriate data input and provide the recommended action depending upon the applicant's rating. The bank manager needs reflect no further, unless there are other issues to consider. Our ethical decision support model will not function in this way. It provides a framework by which the user can think about the ethical issues arising *at the same time* as the technical evaluation takes place. It is designed as an integrated model so that neither technical values nor ethical values are marginalised. Business Ethics has created a lot of material on the subject of company codes.[20] John Donaldson accepts that codes can inform people of what is expected of them, give guidance as to how it is to be done, express values that many participants aspire to and raise standards. He also lists a variety of limitations.[21] But can

we take this as the whole story? Kant's writings illustrate the importance of the will in connection with moral action.

> "Moderation in emotions and passions, self-control, and calm deliberation not only are good in many respects but even seem to constitute a part of the inner worth of the person. But ... they are far from being good without qualification. For, without the principles of a good will, they can become extremely bad, and the coolness of the villain makes him not only far more dangerous but also more directly abominable in our eyes than he would have seemed without it." [22]

For Kant:

> "The good will is not good because of what it effects or accomplishes or because of its adequacy to achieve some proposed end; it is good only because of its willing, i.e., it is good of itself." [23]

This draws attention to a very important matter for business ethicists. The task involved is more than imparting information or convincing the intellect. It involves persuading and moving the will to *voluntarily* perform that which is perceived as good.

This may be illustrated by reference to codes of ethics in societies and corporations. Only the individual can internalize values presented, accepting or rejecting them. Company codes cannot create moral behaviour. They are moral statements that have the endorsement and authority of the CEO, the board of directors or the ethics committee but even the attachment of sanctions will not create positive moral behaviour, but only compliance. It is the presentation of values and their "acceptance" by the individual that produces moral conduct. Moral values may be presented in various ways including the interactive where a writer presents the *evidence* and looks to the reader to draw relevant *conclusions*. This method maximizes reader autonomy in moral thinking but it is not incompatible with rigorous thought. Its aim is to give the reader the opportunity to see for one's self implications, correspondences and necessary actions and thus encourage commitment to the values. Borrowers and lenders will not be impressed by a situation where their world has been forced into some alien framework, but if they perceive that there is a natural correspondence between the stewardship values and their world, then they will be prepared to test their practice by it. The following chapters seek to persuade the reader of the benefits of adopting the stewardship concept in relation to operational debt management.

Notes to Introduction

1. R. Edward Freeman (ed.) *Business Ethics: The State of the Art*, p.7. This book covers:
   (a) societal level considerations,
   (b) stakeholder level questions
   (c) internal policy questions affecting the relationship between the organization and its management and other employees,
   and
   (d) human level questions such as boss-subordinate, peer, network and other relationships.(pp.4-6)
2. T. Lucey, *Management Accounting*, pp.1-2.
3. Alan Pizzey, *Accounting and Finance: a firm foundation,* p.4.
4. Colin Rickwood and Andrew Thomas, *An Introduction to Financial Accounting,* p.1.
5. A.H. Millichamp, *Foundation Accounting*, p.2.
6. Richard L. Daft, *Organization Theory and Design*, p.490.
7. J.M. Samuels and F.M. Wilkes, *Management of Company Finance,* p.6.
8. Tom Cannon, *Corporate Responsibility*, p.71.
9. See Martin Sheard (1992) 'New problems for lenders', *Banking World*, September, , p.21.
   For practical guides to environmental stewardship see Ruth Hillary, *The Eco-Management and Audit Scheme: a Practical Guide* and Christine Hemming *Business Success from Seizing the Environmental Initiative.*
10. John Donaldson, *Key Issues in Business Ethics*, pp. 48-49.
11. Kolb, Rubin and McIntyre (1979) *Organization Psychology: An Experiential Approach*, Englewood Cliffs, NJ: Prentice-Hall Inc., discussed in John Gill and Phil Johnson, *Research Methods for Managers*, pp.24-25.
12. Daniel R. Gilbert, Jr. *The Twilight of Corporate Strategy: A Comparative Ethical Critique.*
13. *Ibid.*, pp. 52-53.
14. For the scope of business ethics see William C. Frederick, James E. Post and Keith Davis, *Business and Society: Corporate Strategy, Public Policy, Ethics* Their book covers :
   (a) The corporation in society.
   (b) The corporation in global society.
   (c) The corporation and public policy.
   (d) The corporation and its response to stakeholders.
   (e) The corporation and social issues.
   For the search for improvement see John Donaldson in his book *Key Issues in Business Ethics*. He utilizes the three principles: pluralism, Golden Rule and autonomy to address the subject of business ethics.
   In his book *Business Ethics: A European Casebook* Donaldson approaches values in business through a wide variety of cases. *Chapter 12* entitled, 'Financial

Services and Normal Values' and *Chapter 13*, entitled, 'The Significance of Scandals in Financial Services' provide an introduction to ethical issues in financial institutions.(pp.224-254)

For use of business ethics literature see John E. Fleming (1987) 'Authorities in Business Ethics', *Journal of Business Ethics* 6, pp.34-51. Fleming made a computer analysis of over 800 references from 57 articles.

The top 5 most-referenced authors were, Milton Friedman, Christopher Stone, Thomas J. Donaldson, Peter French, Alasdair MacIntyre. The list also includes Norman E. Bowie (11th), Peter F. Drucker (15th), John Rawls (18th), Theodore Levitt (26th) and Clarence C. Walton (29th).

The most-referenced article was Steven N. Brenner and Earl A. Molander 'Is the Ethics of Business Changing?' *Harvard Business Review*, 55, no. 1.

The top four most-referenced collections were:

Tom L. Beauchamp and Norman E. Bowie (eds.) *Ethical Theory and Business.*
Thomas Donaldson and Patricia Werhane (eds.) *Ethical Issues in Business.*
Richard T. DeGeorge and Joseph A. Pilcher (eds.) *Ethics, Free Enterprise, and Public Policy.*
Clarence C. Walton (ed.) *The Ethics of Corporate Conduct.*

15. Daniel R. Gilbert, Jr. *op. cit.*
16. There is a range of recent literature that addresses the problem of debt from an economic or social point of view. Such works are primarily concerned with the dimension of the problem, its historical context, causes and remedies. They may make incidental references to moral considerations, but that is not their primary focus.

    A recent historical work is Hugh Barty-King, *The Worst Poverty: A History of Debt and Debtors*. This provides a rare insight into the social impact of debt since Norman times.

    An economic commentary on the debt crisis is provided by Tim Congdon *The Debt Threat: The Dangers of High Real Interest Rates for the World Economy.* His work raises a number of issues relevant to the ethics of debt management.

    There is literature assessing the social impact of debt, mainly by empirical research.

    (1) Janet Ford, *The Indebted Society: Credit and Default in the 1980's* is sociological in its orientation and primarily focuses upon consumer debt. It is of particular value in connection with the types of problem encountered by the consumer and the causes of those problems.
    (2) Richard Berthoud, *Credit, Debt and Poverty* is an important survey report providing statistical information and comment.
    (3) The National Consumer Council *Credit and Debt: The Consumer Interest* report addresses problems arising from the marketing of credit and makes recommendations for improvements.
    (4) John McQueen, *What to do when Someone has Debt Problems: A Practical Survival Guide*. McQueen is the secretary of the *Bankruptcy Association* and has written this book in a popular style. It is packed with practical suggestions

arising from years of experience in debt counselling. It provides a good practical perspective.

(5) Geraint Howells, Ian Crow and Maureen Moroney, *Aspects of Credit and Debt* covers choices available to poorer consumers, the responsibilities of lenders, borrowers and regulators and credit unions.

The *Jack Report* known officially as *Banking Services: Law and Practice - Report by the Review Committee* has 433 pages dealing with a wide range of bank services issues. *Chapter Eleven* on plastic cards and *Chapter Sixteen* on standards of banking practice have some relevance to lending but the report does not address borrowing and lending directly.

17. See, for example, Paul Vallely, *Bad Samaritans, First World Ethics and Third World Debt* and Ngaire Woods, *Ethics and Interests in the International Political Economy, the Management of the Mexican Debt, 1982-1989.*
18. Andrew Hartropp (ed.), *Families in Debt: The Nature, Causes and Effects of Debt Problems and Policy Proposals for Their Alleviation.*
19. James J. Lynch, *Ethical Banking: Surviving in an age of Default.*
20. See for example: Sheena Carmichael and John Drummond, *Good Business: A Guide to Corporate Responsibility and Business Ethics*; John Donaldson, *Key Issues in Business Ethics* and *Business Ethics: A European Casebook*; and Tom Cannon, *Corporate Responsibility.*
21. John Donaldson, *Business Ethics: A European Casebook*, pp.79-80.
22. Immanuel Kant, *Critique of Practical Reason: And Other Writings in Moral Philosophy*, pp.55-56.
23. *Ibid.*, p.56.

CHAPTER 1

THE CONTEXT OF STEWARDSHIP: THE DEBT PROBLEM

1.1. Historical Context

1.1.1. USURY AND SCRIPTURE

Usury has had a long history. It appears in the Old Testament as an important social issue. Understanding usury as exorbitant interest can cloud the subject. This was not necessarily the Hebrew usage, although the word *neh'-she'ch* (or *neshekh)* has its basis in the verb "to bite".[1] This etymology perhaps implies a method of imposing interest in advance so that the borrower agreed to pay back a sum greater than the amount advanced.[2] In Psalm 109, the term *nah-shah* is used with the flavour of extortion [3], but the fundamental idea of the word is "giving time for payment". We should not, therefore, approach the usury debate with the preconception that the Old Testament is negative towards interest in itself. The situation is more complex. Things which may be perfectly legitimate in themselves can become unacceptable as a result of circumstances or extremes. The taking of interest was legitimate in the case of the stranger.[4] It was not, however, permitted to charge interest to an Israelite brother.[5] Nehemiah considered it reprehensible that the rulers should take advantage of their poor brethren by way of interest.[6] Their responsibility was to help without expectation of personal gain.[7] Although the Old Testament does not condemn interest in itself, it is not complacent about debt and its dangers. This is consistent with New Testament teaching. Although in the epistles the word *opheilo* predominantly emphasizes the positive sense of responsibility that arises in consequence of belonging to Christ, in the Gospels it has, in the main, the negative meaning of being in debt with the obligation and uncomfortable consequences involved.[8] Debt is best avoided but usury can be mentioned without negative connotations.[9]

A negative view of interest arose through the influence of Aristotle's view of the barren nature of money leading to the condemnation of interest at the Third Lateran Council. Money was regarded merely as a medium of exchange and the concept of the just price required that the buyer made an equivalent return to the seller. In a loan contract, therefore, the obligation of the borrower was to return the amount borrowed, that is, the principal without interest.[10]

To avoid reduced economic activity ways round the prohibition had to be found. R.H. Preston explains common justifications for payment:

(a) *damnum energens* where the lender "suffered economic damage by lending"

(b) *lucrum cessans* "compensation for surrendering the possibility of gain"

(c) *societas* where "the partner retains control of his own money"[11]

We see in these justifications a measure of economic awareness concerning the costs of lending, including in the case of *lucrum cessans*, the opportunity cost.

Sleeman identifies reasons for *interest* in modern economic thinking: the greater productive power of resources used for investment, the need to give an incentive to forgo consumption, the need for a price to allocate capital between competing uses and the benefit of a market rate of interest to evaluate the worth of an investment.[12] It is clear from these suggestions that interest is seen as having a role in economic dynamics. It draws funds into production resulting in higher outputs and living standards. It serves as an indicator for measurement and comparison. The place of interest has been challenged in modern times.

Paul Mills develops seven consequences of permitting interest: (i) The unjust and destabilizing allocation of returns between finance users and suppliers. (ii) The misallocation of finance to the safest borrowers rather than to the most productive. (iii) A propensity to finance speculation in assets. (iv) An inherently unstable banking system that requires government guarantees. (v) A 'short-termist' investment strategy. (vi) The concentration of wealth into fewer hands. (vii) A rapid flow of financial capital across countries.[13] It is debatable whether interest should be the focus of a critic of the market economy. The consequences identified by Mills are not solely or primarily dependent upon permitting interest. If the allocation of finance is taken as an example, the existence of interest and the possibility of including a risk premium in it, could provide lenders with an *incentive* to lend to more risky but productive borrowers. If the instability of the banking system is taken as a second example, it can be shown that the instability does not arise as a result of borrowing and lending where interest is paid. It is rather a consequence of the lending of fictitious funds. Before the advent of fractional reserves, banking was not *inherently* unstable.

Despite our disagreement with Mills about interest his following statement is highly significant:

> "a number of our economic ills can be ascribed to our economic system being reliant upon interest-based debts rather than non-interest financial arrangements."[14]

The key to the interest debate is here. The real need is to address the ethics of the financial *system.* A focus upon interest can be misleading. The Old Testament regulations had already pointed to this. Neither lending nor interest were presented as evils in themselves but the requirement of physical assets as security produced an economic system very different from our own. It is not the abolition of interest that should be the focus of our attention. It is the financial system that requires ethical scrutiny to determine the changes that need to be made in order to remove the economic ills to which Mills has drawn attention.

1.1.2. THE HISTORICAL CONTEXT

The history of debt in Britain is a long one. Hugh Barty-King traces it in his book *The Worst Poverty: A History of Debt and Debtors*. As early as the Magna Carta (1216) debt was an issue. The charter stated,

> "Neither the king nor his bailiffs should seize any land or rent for any debt so long as the present goods and chattels of the debtors should suffice to pay the debt, and the debtor himself be ready thence to satisfy it..."[15]

Canon Law ruled that usury was a sin and those who practised it were debarred from the Eucharist. A statute in 1542 provided for imprisonment of debtors until they made repayment. By 1570 a procedure for bankruptcy was established which lasted centuries. The definition of "a bankrupt" was widened in 1604. In the seventeenth century imprisonment for debt was a severe business.

> "Once in prison the law did not care whether the person had the means to pay what he owed, or not. It was not concerned whether or not he was able to keep himself in life."[16]

Moses Pitt drew attention to the plight of imprisoned debtors in his book, *The Cry of the Oppressed* (1692) but it did not stir the authorities to action.[17] Prisoners in the eighteenth century might be doctors or lawyers.[18] Daniel Defoe's marine insurance losses resulted in him being made bankrupt for £17,000. Neither was the government immune to debt. In 1715 the National Debt was around £37 million.

There is a long list of efforts made for the relief of debtors. When James Oglethorpe MP was horrified at the death of a friend in Fleet prison, a House of Commons Commission of Inquiry into debtors' prisons followed which resulted in the passing of the Debtors Act in 1730. The Act promised relief for debtors willing to satisfy creditors to the fullest extent possible.[19] Technicalities meant that this humanitarian venture could be easily nullified. Additionally it was already being recognised that lenders and sellers could not afford to be sentimental. Customers might perjure themselves to secure bankruptcy and evade paying back what was owed. Oliver Goldsmith's play *The Good-Natured Man* provided a controversial commentary on the borrowing and lending of the times.[20] Practical action included James Neild's *Society for the Relief of Persons Confined for Small Debts* which was founded in 1772.[21] The social conscience of the nation was responsive. Publications such as William Dodd's *An Account of the Rise, Progress and Present State of the Society for the Discharge and Relief of Persons Imprisoned for Small Debts* sustained public interest.[22] Concern for confined debtors continued and in 1788 a pamphlet entitled "New Candid and Practical Thoughts on the Law of Imprisonment for Debt" appeared. Although social action continued, it was not until 1869 that the passing of the *Debtors Act* and the *Bankruptcy Act* ensured that insolvents could discharge non-fraudulent debts through personal bankruptcy.[23] In 1870 the last release of debtors from the Whitecross Street Prison took place.[24]

Even those involved in philanthropy were to find themselves with financial worries. Lord Shaftesbury, following a variety of financial troubles, gave an assessment of his own errors:

> "I acted upon feeling and trusted to the conclusions of my imagination, not realizing the fact that there is no promise of miracles to supply what might be done by common-sense, and that mere warmth of heart is a very deceptive guide in the details of life".[25]

This brief summary of the social impact of debt introduces us to the kinds of concerns that arise: heavy-handed recovery methods, debt dependence, charging interest, ability to repay and make a new start, bankruptcy through business losses, the growing National Debt and relief for debtors.

1.1.3. THE PRESENT DAY

By the turn of the twentieth century the idea of credit had come into vogue. Debt could lose its stigma. "Credit was the solvent that oiled the wheels of many ... retailers' businesses" and having an account became a status symbol. (26) Credit facilities came to be expected and credit assessment grew in importance. In the post-war period home-ownership became a desirable family objective. Attitudes to debt were changing. In his 1986 Rathbone Memorial lecture, Sir Gordon Borrie explained this new orthodoxy that:

> "everyone should be free to obtain as much credit as he could get, on the easiest terms available on the market. But that was not to say that people had a right to it, that they were entitled to it."

The consumer credit provider side of the market emphasized, however, the need for responsible lending.(27)

TABLE 1.1. Consumer Credit - Outstanding at end of period in £ million [28]

	1993	1994	1995
Credit cards	10,654	11,910	13,838
Other	42,660	46,287	51,281
Total credit	53,314	58,197	65,119

The twentieth century replaced "debt" by "credit" as credit promotion became an integral part of business expansion and involvement with debt became more widespread as interest grew in owner-occupier house purchase. The history of debt shows that it can be troublesome for individuals, businesses and governments if it is not properly managed. But it also shows that there is an attraction in debt for the borrower, trader and lender. Borrowers enjoy present satisfaction while deferring payment. Traders secure immediate cash flows. Lenders secure a return on capital through taking risks providing finance. Because debt can be so advantageous to all parties, there is a danger that credit-granting practices become less rigorous than they should be.

The economics of debt is reflected in the appropriate statistics. *TABLE 1.1* shows the levels of consumer credit in recent years. The figures reveal the upward trend in the total consumer credit outstanding both on credit cards and from other sources and indicate an average level of consumer credit around £1,000 per head of population. The figure would rise to £8,000 if mortgages were included. Debt raises many issues beyond the economic sphere. The personal and social consequences of serious debt make it plain that debt must be viewed in a wider context than that usually accorded to commercial transactions. At the time of incurring debt material wants, finance and legal documentation are primary considerations but when repayment difficulties are experienced a host of unforeseen difficulties arise. Recent research conducted by the *Jubilee Centre* revealed that personal debt could result in mental problems, social withdrawal, physical ailments, marital problems and family problems.(29) One hundred years ago ill health arose from sanitary problems, now debt is a serious contributory factor.

1.2. Moral Concerns and the Debt Problem

1.2.1. ECONOMICS

Capitalism

We find ourselves within a capitalist system and have to decide what can be done about it. Capitalism means different things to different people. The following definition is given in *A New Dictionary of Christian Ethics*:

> "...capitalism is that economic system in which the price mechanism, working through supply and demand in workably competitive markets, provides the dominant mode of making economic decisions about what is produced, how it is produced, and in what way it is distributed." [30]

The marketplace is a fact of life in society and, as dispensing with the market place cannot solve our problems, it is evident that the participants within a market economy must have appropriate values to set moral limits. In his book *Ethics in Modern Management*, Gerald J. Williams mentions deception, misrepresentation of product content, and price fixing as undermining the market mechanism.[31] Such behaviour, although not an argument for abolishing the market mechanism, is certainly an argument for recovering appropriate values in business. One writer goes so far as to say that:

> "if the capitalist society as it is known today is not characterized by values, institutions, or behavior with a high level of ethical content, no blame should be attached to the resource allocation mechanisms, or to private property, or to the free enterprise economy, or in general, to the market system."[32]

Even those who disagree most strongly with this comment would be unlikely to dispute that, *whatever* the system, people's dealing with one another is still fundamental to satisfactory operation.

Although the competitive market is central in the definition of capitalism there are several associated ideas: (1) the rule of law upholding economic freedoms, (2) private ownership of the means of production and (3) profit motivation. Emphasis on an impersonal price mechanism and on profit motivation make capitalism an easy target for criticisms of heartlessness, environmental exploitation and greed but callousness and greed can operate in any system, just as kindness and generosity can be conceived of as compatible with the market place and profit-making. Recent work by Werhane has advanced the case that Adam Smith's central thesis was not 'self-interested utility maximization' but the virtue justice.[33] Certainly it is possible to conceive of entrepreneurial activity and market exchanges within a context of justice. The Old Testament prophets exhort Israel to such a situation where each pursues his calling within a context of fair markets.

The case against capitalism has been advanced on the basis that what is needed is the abolition of private property. Such a presupposition moves from economics to more fundamental matters. Critics of capitalism have recognized this. Noonan, for example, insists that the evils of capitalism are not due to the *ownership* of private property but the *abuse* of such ownership. It is not the institution itself but the distorted interpretation of it that has the destabilizing effect. One problem is that the moral limits get overlooked. So, for example, there is a need to recognize that Divine sovereignty limits acquisition as the Divine Being

> "... is the first and absolute owner of all things, and this dominion over created things He can never relinquish. Our ownership of them is therefore secondary, and that of stewards only."[34]

The introduction of moral limits into the debate is of importance because the economic system must be considered in the light of moral values as well as technical values. When this is recognized it becomes clear that market economics do not have to be synonymous with the commercialization of life or the domination of society by monopolies. The whole issue of decision-making involves more than the quantitative, and there is no guarantee that public bureaucratic decision-making will produce better qualitative results than private decision-making. Labour hoarding by nationalized industries in the interests of employment provides a short-term remedy but forces up prices and fosters longer-term problems. Critics of private sector short-termism can thus end up with their own brand of short-termism. The need is for values that recognize economic reality and social and environmental responsibility.

In this connection the debate about the link between religion and capitalism is important. Max Weber sought to shed light on the link between religion and economic activity in *The Protestant Ethic and the Spirit of Capitalism.*[35] This analysis was continued by Professor R.H. Tawney in his book *Religion and the rise of Capitalism.*[36] Both writers recognized that the Protestant concept of "calling" had released tremendous energy in economic activity and neither writer seems to have regarded the Protestant work ethic as synonymous with *laissez-faire* capitalism.

A major concern of Professor Tawney was the secularization of the economic system. The sanction for human affairs was divorced from the supernatural, society is viewed individualistically and mechanistically and moral expediency had replaced generally acceptable moral standards.[37] Developments in the half century since Tawney's analysis have done nothing to invalidate his concerns about an unspiritual and de-humanized economic system. Contemporary futurists, for example, are making criticisms about a culture driven by instrumental rationality, a reductionist world view and a lack of spiritual values.[38] The search for new purposes and meanings, should not be seen as an opportunity to disregard the Protestant work ethic on the basis of some supposed identity with contemporary capitalism. In several respects the former has more in common with the futurists, than with the latter. Futurism is concerned with the importance of addressing the future by thinking into the future. Hazel Henderson explains how it seeks to construct alternative futures by:

> (a) The use of "What if?" scenarios rather than extrapolating trends.

(b) Identifying *Preferred Futures* to accompany economic forecasts.
(c) The assumption of fundamental change rather than equilibrium.
(d) A proactive rather than reactive intent.
(e) Non-linear rather than linear reasoning.
(f) An organic rather than inorganic system model.[39]

The Protestant Work Ethic, strong for 300 years from the sixteenth century, has more in common in some respects with modern Futurism than Capitalism. All three agree with:

(1) A place for innovation, diligence, creativity, and goal-oriented reasoning.
(2) The recognition of stewardship as a relevant relationship in economic activity.
(3) The need for a value framework for work and economic activity.

Additionally, capitalism and the Protestant work ethic both accept the link between productive work and prosperity, the legitimacy of personal wealth and the importance of investment for future effectiveness.

The Protestant work ethic and Futurism criticize secular capitalism's:

(1) pre-occupation with presupposed fixed mechanistic laws;
(2) replacing *Ethical Man,* a moral agent, with a rationalistic *Economic Man*;
(3) regarding economic decisions as value-free; and
(4) excessive emphasis on technical values instead of moral and human values.

Table 1.2 shows the ways in which the Protestant work ethic suggests a different approach and outcome to that of secular capitalism.

TABLE 1.2 Comparison of the Protestant Work Ethic and Secular Capitalism

	Protestant Work Ethic	*Secular Capitalism*
GOALS:		
Economic Growth	An effect of service and measure of its effectiveness	An end in itself
MOTIVE:		
Social benefit by	Mutual benefit, co-operation	Self-interest, competition
STANDARDS:		
Work	Service	Means to prosperity
Social dimension	"Man in community"	Individualistic
Limited liability	More liability for others	Financial cornerstone
Debt	Debt-averse	Debt-dependent:
METHODOLOGY:		
	Regards the future as neither fortuitous nor wholly predictable	Prediction using rationalistic assumptions about human behaviour.

We have seen in this section that capitalism is an economic system based on the price mechanism, that transactions in the marketplace are affected by the values of the participants and that the abolition of the price mechanism or of private property are no guarantee of an improved society. In any economic system moral values are as important as technical values. The Weber-Tawney debate focused attention upon the problem of secularization in economics and the emphasis upon individualism and mechanisms. The Protestant Work Ethic is consequently far from redundant and anticipated Futurism's criticisms of capitalism. All three positions agree on the importance of diligence, innovation, creativity, goal-orientated reasoning and stewardship. But Futurism and the Protestant Work Ethic unite against secular capitalism in rejecting mechanistic rationalism, economic man, assumed value-free economic behaviour and excessive emphasis upon technical values. Fundamental change is essential and the Protestant Work Ethic shows that goals, motives, standards and methodology are all involved.

Rationality

Rationality Required. The Concise Dictionary of Christian Ethics defines the economic principle in a way that reveals its central importance for effective resource management.

> "It may be said of the so-called economic principle (rational principle) which underlies every economic system that it rests fundamentally on the aim of winning the greatest possible success (output) with given means, or of winning a given success with the smallest possible input."[40]

Looked at in this way it can be seen that there is an indispensability involved. In times when the importance of conservation and sustainability are recognized it can be seen that a rational basis for getting the best out of available resources is necessary. Note firstly, that this principle is one thing considered as a technique and another if elevated to a total morality; and note secondly, that the rational principle addresses the ideal, to secure the best output from limited resources, not the actuality, which for a variety of reasons might be different. There is consequently a balance to be maintained. Efficiency of resource allocation is a principle that should not displace social norms such as fairness, justice and social values, but equally, social values *cannot* change the necessity for efficiency and should not displace it. There is a place for economics and ethics. Amartya Sen has argued that the methodology of positive economics requires balancing by normative economics [41] but this is far different from assuming that somehow it is possible to get along without the disciplines of the market. We can neither assume that the market will always result in fair outcomes [42] nor that all participants will act fairly.[43] Even an efficient market system can be defeated by questionable ethics.

Rationality Challenged. The idea of a market based upon rational choice has been challenged suggesting that it is unrealistic to expect or to encourage "rational" behaviour. A major criticism of economic theory has been its dependence on the assumptions that producers and consumers will act rationally, when the reality is that there is a high degree of unpredictability in the behaviour of both. So Jon Elster argues:

> "The fallacy of ideal models of rational decision making is the assumption upon which they are based. They assume that tastes are absolute, relevant to the issues, consistent, and precise, when in fact none of these conditions are guaranteed."[44]

The second sentence accurately states the problem of the level of confidence that can be placed in outcomes predicted by economic modelling by private sector firms or government. But being incomplete or provisional does not invalidate a technical model. It only becomes a fallacy if the partial account is taken to be the full account.

Rationality Defended. Sharon H. Garrison makes the point that market theory does not necessitate that all participants act rationally but only that the majority do.

> "Whether all investors in a market act rationally might be subject to question. However, if a sufficient number of traders are rational, the forces of supply and demand will dictate that actions of irrational participants will be offset and that goods should trade at a price reflecting their inherent value.[45]

Garrison accepts that rational behaviour does not have to be completely comprehensive for the price mechanism to be effective but this would still require some specification of the nature of rational behaviour. This is a significant aspect, not only of the debate about economic theory and market effectiveness, but also of the role of rationality in economic action.

The Nature of Rationality. Sen has distinguished two main methods of construing rationality in economics. In one case rationality is seen as

> "internal *consistency of choice", always acting in the same way, and in the other "maximization of self-interest".*[46]

This distinction is by no means unimportant, because if we move to the former because of ethical dislike of the latter, we are then compelled to face the fact that rationality is about more than consistency. Sen puts it as follows:

> "If a person does exactly the opposite of what would help achieving what he or she would want to achieve, and does this with flawless internal consistency (always choosing exactly the opposite of what will enhance the occurrence of things he or she wants and values), the person can

> scarcely be seen as rational..."[47]

To the normal thinking person rationality evidently involves some correspondence between *ends* and the choice of *means* and requires that the latter be appropriate to accomplish the former.

A Reasoned Approach to Debt Management. Encouraging a reasoned approach to debt management is different from a rationalistic decision process that ignores the human dimension of business activity. Stewardship is a form of reasoned decision, but does not depend on rationalistic modelling and Robert E. Lane explains that it is possible to have the former without dependence upon the market's rational decision making model. He cites by way of examples, scientific decision making and legal decision making.[48] This point can be illustrated by reference to the chaos theory of management. The latter is particularly pertinent to the debt default situation. It emphasizes the existence of turbulence and the importance of managers being able to operate in the boundary region between order and disorder. Dr. David Parker explains the situation as follows.

> "In chaotic systems links between cause and effect are difficult to identify and therefore it may not be possible for management to operate on specific levers for change. ... Each firm must develop its own distinct strategy, a process in which 'generic strategies', ... which by implication apply to all industries and at all times, provide broad guidance only..."[49]

Thus even when things are beyond control, and in the case of debt, have gone very badly wrong, there is still a place for strategy and thoughtful, reasoned action and, as we will show later, stewardship helps to regain control. This is important because actions influence outcomes even in uncontrollable situations. Parker further states:

> "The existence of complexity and unpredictability in cause and effect relationships does not mean ... that management cannot influence outcomes by its actions ... Management must plan how much to produce now and in the immediate future..."[50]

There is thus a place for rational response as long as the limitations with respect to the future are recognized. By deepening their understanding of what is necessary the participants will be able to make a more effective contribution. This will be true of borrowers and lenders. In developing my argument it is not assumed that there will not be instances of irrational behaviour in connection with debt management. It is assumed, however, that the normal individual is capable of rational thought, most of the time, and that a reasoned approach to problems will usually produce a better outcome than irrational behaviour.

Government Policy and Sound Money

Parliament secured control of national finances towards the end of the seventeenth century. The formation of the Bank of England just before the end of the century was linked to a loan to the government of £1.2 million. The two following centuries are generally regarded as the period of *laissez-faire*. The use of this term needs to be viewed with some caution. It has been shown that the term was not used once in the 1500 pages of *Hansard* covering the Free Trade debate in 1846. In his book *The English Ideology: Studies in the Language of Victorian Politics*, George Watson states of *laissez-faire:*

> "No political party in nineteenth-century England can be shown to have believed in it or to have attempted to practise it..."[51]

The Manchester School did indeed strongly advocate government non-interference but Asa Briggs maintains that even this school was divided.[52] The nineteenth century saw increasing Government involvement in education, health and highways. The National Debt was little reduced from the £800 million to which it had grown by the beginning of the period. However, the impact of the Government upon the economy was minor compared to what was to follow in the next century.

Intervention received a massive impetus through Lord Keynes. The strengths of his position were that he focused upon investment as the key motor of economic prosperity and recognized that savings and investment are governed by different judgements about the future.[53] The Keynesian analysis suggested that a timely bolstering of deficient effective demand could reduce unemployment. Increased public expenditure was seen as one method of compensatory action. Milton Friedman accepts that although fiscal policy is significant in determining the public expenditure proportion of national income and the distribution of that expenditure, it is not inflationary if the expenditure is financed by taxation. If, however, the government spending is financed by printing currency or by credit creation the public expenditure will be inflationary.[54] Friedman comments:

> "...full adjustment to monetary disturbances takes a very long time and affects many economic magnitudes."[55]

If, as we have seen, sustainable development requires rational decision-making based upon reliable financial data, such long-term effects are a serious matter. Their dimension and duration can upset the best individual plans and individual stewardship must be supported by responsible public stewardship. Hayek has also expressed concern about the impact of using the money supply to achieve political goals expressing the opinion that:

> "...to use the control of the supply of money as an instrument for achieving particular ends destroys the equilibrating operation of the price mechanism..."[56]

These views have support from the growing concern among social ethics writers that undermining the soundness of the currency can have long-term detrimental effects. Ronald H. Preston, for example, while favouring the retention of government control of the money supply to provide for social policies, recognizes the injustice caused by varying money's purchasing power.[57] The issue cannot be addressed solely in terms of justice for the socially disadvantaged. Once an inflationary spiral is triggered off by monetary expansion it impacts differently on different firms. A firm may have borrowed to finance profitable projects, it may be operating efficiently, but if it faces cost increases as a result of inflation, at a time when market conditions will not permit price rises at the same rate, its profit margin will be eroded and it will be forced into a loss-making and debt default position. The issue is not merely about the re-distribution of income. It is also about the destruction of productive business firms. What may be seen as the alleviation of hardship on the one hand, can create bad debts on the other, even when borrowers have been responsible and businesses have been efficient.

1.2.2. BUSINESS

Organizations

Complexity. The complexity and inter-dependence of modern society necessitate the organization of the activity of work. As technology advances, specialization increases and products become more sophisticated. Organizations must develop systems and procedures to govern internal and external relationships and as they do so they are influenced by managerial and employee values, economic realities, public opinion, legislation and international agreements. The organization must clarify its purpose, identify the responsibilities that it has to internal and external parties and its expectations of them.

Key Areas. This gives rise to three key areas of moral concern: (a) the definition of the *corporate purpose*, (b) the recognition of *constituency interests*, and (c) the specification of *codes of conduct*. The first of these is considered in this section and the remaining two are considered in the following sections.

Unacceptable Bias. The diversity of viewpoint about what the corporate purpose should be focuses attention upon the importance of values within the organization. Richardson and Curven have argued that free-market doctrine, which we have recognized as pre-occupied with technical values, tends to promote a biased view of organizational life. They identify six relevant principles:

(a) the principle of the organization as the predominant benefit creator,
(b) the principle of survival of the fittest through competition,
(c) the principle of selfishness or self-interest,
(d) the principle of profit maximization to the exclusion of social factors,
(e) the principle of short-termism. [58]

Richardson and Curven argue that such deep-seated beliefs promoted by contemporary free market capitalism contribute to a crisis-ridden society.

Remedies. There is a spectrum of opinion among those who want to see greater social contribution from corporations. Some want more corporate philanthropy, some community involvement and others a corporatism within which the company provides a linkage between government and the individual. But as John Atherton has clearly pointed out there is a deeper level of analysis of purpose needed. He comments as follows:

> "...although effective long term economic performance is generally accepted as more than a narrowly economic issue, it can never be achieved in isolation from the pressures for efficiency in a global market place. General Motors, the largest TNC in the world, was managed very effectively in the 1970s, in relation to shared goals and flexibility. However, the commitment to fairness and effectiveness was no protection, ultimately, from the competition of more efficient Japanese corporations. Effective corporate mission purposes have always to balance efficiency and fairness."[59]

It is this balance of *efficiency* and *fairness* that stewardship is able to address by providing an ethical framework addressing both resource and relationship dimensions of organizational activity. Ethical investment has shown that shareholders are by no means averse to a variety of value issues in the conduct of business. The central issue is that there must be a core value structure that is applicable to all firms and which balances the need for profitability and the need for fair dealing with all stakeholders. The former is a matter of survival in the face of economic realities, and the latter is a matter of facing up to what is just and fair. Both of these are moral concerns. Poor economic stewardship wastes resources and imperils jobs, so irresponsible management results in greater hardship than a management too focused on profit maximization but which stays in business. The need is not to *add on* social roles to the economic role of business organizations, but to *build in* generally accepted core values so that firms function in a socially responsible way. Stewardship is concerned with core values and I will show later how stewardship can answer to this need.

Constituencies and Stakeholders

Stakeholder Interests. Business ethics literature looks at company relationships in stakeholder terms Evan and Freeman presented a stakeholder theory of corporate operation to emphasize that shareholders are only one of the parties involved. It is now common to see lists of such stakeholders in management texts. Certo and Peter identify suppliers, lenders, government agencies, political groups, employees, consumers, competitors and the local community in addition to stockholders.[60] Stewardship is also concerned with constituencies other than internal stakeholders. The special role of management, according to Evan and Freeman, is to look after the corporation's health balancing multiple claims of the conflicting stakeholders. No stakeholder group has primacy and management must keep relationships in balance.[61] Banks must balance the interests of shareholders, depositors and borrowers.

The Managerial Task. Kuhn and Shriver prefer to use the term "constituencies". They fear that managers accept the term "stakeholder" but refuse its substance. They believe that the special trustee role of balancing interests does not promote managerial accountability, only a self-defined sense of responsibility. [62] Their perception of the unique managerial role:

> "is not in providing a particular goal or set of values for the firm, such as assuring survival, but rather in fulfilling a necessary *function:* co-ordinating all the various demands, negotiations, and settlements so that the outflows of resources match inflows."[63]

According to this view the policy and direction of the organization is determined by the complex of negotiations among the various constituencies and as managers recognize the demands of various groups and accept their input and contributions this leads to effective operation. In their view it is too much to expect managers alone to cope with the many social, economic and political problems arising on a daily basis.[64] We agree with these authors that resource inflows and outflows must match, and that constituency interests must be addressed, but consultation and participation are time consuming and expensive and resources would be saved by employing an ethical decision model incorporating economic matching and values relevant to ethical relationships.

Uniting Around Core Values. Whatever the level of involvement of the different constituencies the economic fundamentals have to be addressed and this necessitates the acceptance of a common framework of relevant values. Sir Adrian Cadbury puts one issue sharply when he explains:

> "Satisfying customers requires companies to compete in the marketplace, and so we cannot opt out of introducing new technology, for example, to preserve jobs."[65]

The failure to be internationally competitive destroys jobs. Faced with shedding 25% of its workforce or going out of business, a firm may not find participation useful, and in the long term it will not serve as a substitute for the dynamic provided by relevant values so that the organization can continuously develop in a way that will maintain its effectiveness in a changing environment to the benefit of all of the constituents.[66] Stewardship is not about being static but about managing change in a value-sensitive way. Values are more than a control mechanism. The right values are engines of improvement and beneficial development. What is needed is acceptance by constituencies of relevant core values. This is not impossible. Christopher J. Cowton, has maintained that it is possible to reflect ethical preferences in investment because "experience shows that people of different basic beliefs or values can discover common ground around particular issues."[67] Such agreement draws in funds but successful operation requires that employee acceptance and behaviour sustain the values.

Business Codes

The *Institute of Business Ethics* research shows that the percentage of large companies with codes has increased from 18% in 1987 to 47% in 1995.[68] Codes are revised to reflect changing circumstances, are general in scope not industry-specific and make reference to sanctions. Corporations consider codes necessary and use them to guide and control employee behaviour. Codes can make clear to employees what is required of them and give guidance about grey areas providing some mechanism to improve standards [69]; but they are no substitute for our stewardship decision model. The ethical content of codes is limited and of a general nature. Enforced morality cannot replace the synergy arising from core values [70] and can be counter-productive encouraging mechanical reasoning and stultifying judgement.[71] The need is for internalized core values to generate cohesive action. Without the dynamism of relevant values organizations remain vulnerable. Our stewardship model provides values that, not only employees, but all financial services industry participants can accept.

1.2.3. PRIVATE INDIVIDUALS

We now move on from the organizational perspective to that of individuals and address the specific difficulty of irrational debtor behaviour.

Personal Debt Problems

The literature about debt expresses concern about consumer credit use, its commercial encouragement, mortgage risks, the impact of debt on women, disconnections and credit card users. Both the seriousness and scope of the debt difficulties are emphasized and the need to tackle the root causes of the problem. We consider at this point the serious problem of irrational response.

Irrational Behaviour

We discussed rationality in relation to capitalism and the market economy and now relate what was said to debtors. The debt situation in general and default in particular can be aggravated by irrational behaviour arising from panic or depression. At such times, even if following medical treatment of the debtor, the money adviser will have to enable the debtor to approach the debt problem in a reasoned way. Predictive modelling cannot answer to this need. The stewardship model developed in this book is not designed to predict consumer or lender behaviour, but to inform ethical action. In accepting the ability of individuals to act rationally, the writer is acknowledging the humanity of borrowers and lenders and providing a comprehensive ethical model consistent with this.

1.3. Technical Aspects of the Morality of Debt

1.3.1. ECONOMIC ISSUES

Debt Dependence and Over-commitment

Financial analysts can see when things are out of control. The presupposition respecting debt is that eventually it should be repaid. If it cannot be repaid the debt is unsustainable. Congdon examines the problem where the interest rate exceeds the growth rate of income of individuals, firms or governments and non-interest expenditure is not reduced. He accepts some upward movement in the debt to income ratio resulting from further information facilitating creditworthiness appraisal, because the security requirements of lenders provide some brake on unsustainable increases.[72] He comments,

> "...there would be a problem if the debt/income ratio was rising and new debt lacked adequate collateral in the form of tangible assets. When debt is similar in value to, or somewhat exceeds, a nation's tangible assets, further increases in debt should not proceed more rapidly than increases in national income."

Congdon refers in this connection to the "financial interrelations ratio" (FIR) calculated by the formula,

$$\text{FIR} = \frac{\text{Financial Assets}}{\text{Tangible Assets}}$$

When the FIR exceeds unity some of the debt will be unmatched by tangible assets. The FIR limits debt expansion, because in even the most sophisticated economies this ratio seldom exceeds 1.5. Congdon concludes that, in industrial countries with an FIR above unity, the growth rate of debt should not be greater than the growth rate of national income if problems are to be avoided.[73] Increases in debt need not result from new lending. Of the borrower facing an interest rate above the growth rate of his income Congdon explains,

> " Unless he makes an effort to service his loan ... the loan increases every year by the ... interest charges. If there is no debt servicing whatever, the growth rate of the debt is the interest rate. It follows that, because the interest rate is above the growth rate of income, debt increases faster than income and the debt/ income ratio rises."[74]

This is the hub of Congdon's argument and he considers it applicable to any economic agent. As interest is paid on accumulating interest debtors are caught in a spiral as "interest on the interest exceeds interest on the principal"[75] and without austerity measures the debt/income ratio rises explosively. Such technical analysis is important but we will introduce moral values and social considerations as well.

Interest Charges

The issue of interest is thus very important for debtors, and, as we have seen, interest charging has been recently challenged in a Jubilee Centre Publication, *Interest in Interest: the Old Testament Ban on Interest and its Implications for Today*.[76] But the morality of interest is not so significant an issue at the operational level for two reasons. Whether interest will be charged or not is at the discretion of loan officers. Even with respect to the interest burden lending officers usually have only a marginal influence on the *rate* of interest charged. When the kinds of problems Congdon deals with arise for debtors as a result of interest charges it is usually related to movements in market rates. The level of interest rates is a complex matter and those who have any awareness of the practical situation know that bank managers are as powerless to influence interest rates as anybody else except within narrow limits. Loan officers may determine the interest margin over base rate in the light of the degree of risk involved but the market rates will determine the level at which the base rate is at any given time. In connection with personal lending the risk premium may be set by the bank Head Office and loan officers apply the advertised rate. In both cases, unless the rate is fixed, the impact of market rates on base rates and lending rates determines the interest repaid. The determination of market interest rates is complex. Jeff Madura identifies three impacts: inflation, government budget deficits and foreign interest rates.[77] This enumeration cautions against a consideration of the level of interest rates independently of the morality of macroeconomics. The issue of interest is a community and global matter and properly belongs to the ethics of macroeconomics and not to the morality of operational lending. What borrowers must take into consideration, however, is the volatility of interest rates and the impact on finances of long-term rate increases resulting from macroeconomic changes.

Lax Advance Control

Congdon raises the question as to why the lender is so misguided as to keep on extending credit when the borrower's position is untenable.[78] He directs attention to two lender attitudes to explain the apparently irrational behaviour. The first is reliance on security. As long as asset values increase, the banker has a comfortable margin of safety in the event of default even if the borrower's ability to service the loan deteriorates. If, however, money values of assets fall quickly, surpluses will disappear. The second attitude is loyalty to customers. Bankers see it as their functions to finance customers through cash flow difficulties and such periods often coincide with times of transient weakness in asset values.[79] Lax control can thus arise from over-confidence or the desire to help or a mixture of both but it can be seen that the dynamics of advance control involve making fine judgements about the extent of the support given and the length of time that such a position can be maintained in the light of the economic circumstances. The bank must make daily judgements, not dissimilar to the borrowing firm's management, about expected revenues and the time of their receipt. The security position will be relevant because, once the margin of safety has disappeared, a wrong judgement about the bank's involvement will involve significant losses. There are other factors. F.W. Pointon gives the example of clearing banks with large networks whose own smaller customers would be adversely affected if the bank stopped supporting a large trading company resulting in its insolvency.[80] The danger of laxity is always present, but the dynamic decision complex of advance control is part of the explanation as are the remoter consequences of adverse action. Such hard technical judgements necessitate a robust ethical framework.

Debt Recovery and Remission

Unsustainable debt is bad for bankers as well as borrowers because the ratio of bad to good loans increases and bad debts increase.[81] All of the ratios are vital because neither party has control of macroeconomic impacts and what looked a sound proposition can soon turn bad as Third World commodity producers faced with interest rate increases above commodity price increases found. In such circumstances there is pressure on lenders to alleviate the debtor's plight. Debt remission is one approach. Third World Debt has been a continuing problem. In *A Fate Worse than Debt* Susan George identifies the following causes: arms purchases, useless projects, cost-inflated projects, conspicuous consumption, capital flight, interest rates and oil prices.[82] Debt forgiveness was one proposal for dealing with the international problem. In 1989 Nicholas Brady proposed debt relief through forgiveness as an alternative to less-developed countries undertaking growth-orientated structural reforms. The 1980s international debt crisis illustrated the need to get the fundamentals of expenditure, income and borrowing right. Shapiro does not see forgiveness as the answer for such countries.

> "Their problems are caused by governments spending too much money they don't have to meet promises they should not make. They create public sector jobs for people to do things they shouldn't do and subsidize companies to produce high priced goods and services. These countries need less government and fewer bureaucratic rules. Debt forgiveness or further capital inflows would just tempt these nations to postpone economic adjustment further."[83]

Not all readers will agree with Shapiro's analysis, but his argument holds good that debt forgiveness and further capital inflows will not help if the economic fundamentals are not addressed. In *The Debt Boomerang: How Third World Debt Harms Us All,* Susan George catalogues the burden of Third World debt on northern taxpayers. She includes the costs arising from official lending which ends up being used by third world governments to meet debt service costs payable to banks; and the inefficient use of public funds in Brady style debt reduction strategies.[84] The dilemma underlines the need for a model that balances economic fundamentals and broader moral and social values in lending.

Unsustainable Budgets

One macroeconomic influence damaging to individual finances has been deficit public budgets. Economic fundamentals are reflected financially in budgeting. When income exceeds expenditure there is a long-term viability. When income persistently falls below expenditure debt problems increase. A major contributory cause of public debt problems is seen by Congdon to lie in the rejection of the necessity to match income and expenditure that he sees as having ethical roots. Public finance has followed the rationalizing tendency of economics, as theories and analysis have increased in sophistication, while the moral undergirding of financial responsibility has declined. This led to a departure from the early twentieth century emphasis on responsibility, prudence and caution that reduced the national debt burden [85] to an acceptance of budget deficits under the impetus of the Great Depression. This approach can be traced through Keynesianism, Kennedy and Johnson and improving USA % growth rates of real GNP in the sixties seemed to confirm the value of deliberate budget imbalance although the impact of Keynesianism on policy-making was not as prolonged and prominent as has at times been suggested.[86] There is no doubt that interventionism was well-intentioned. Keynes *The General Theory of Employment, Interest and Money* [87] was published in 1936 at a time of high unemployment and sought to provide a solution by compensating for a shortfall in effective demand arising from weak private consumption or private investment. The recommended remedy was to bolster effective demand by filling the gap between total income and total consumption by increased public expenditure based on deficit fiscal policies that increased the national debt. As long as such a policy is considered in isolation from the position of individual investors, individual companies and individual borrowers the full implications of this technical model are not disclosed.

Congdon explains how investor awareness that pragmatic governments are not committed to budgetary balance, maintaining a flexible response to competing policy objectives, causes savers to anticipate that inflation will be tolerated,[88] and,

> "... the looser is the government's commitment to sound money, the greater is the weakening of savers' confidence, and the weaker is savers' confidence the higher are interest rates."[89]

Through rising interest rates the government budget deficit impacts upon the budget arrangements of individuals and firms. The macro movements within an economy dwarf their micro decisions. Thus, even though at the time of borrowing a domestic budget may have been sustainable, the increase in interest rates soon renders it otherwise. We see, therefore, that there is a linkage between issues of trust between savers and government and the economics of debtor households, in consequence of which, reasonable levels of caution on the part of a borrower can be swept aside by government imprudence while pursuing social objectives.

Financial Measurement

R.H. Preston describes the impact of hyper-inflation on Brazil where

> "Firms could make more money by financial speculation than by investment" and where government finances were destroyed "as tax revenues became worth ever less as they came in, so that very high interest rates were charged for borrowing."[90]

The control of inflation is fundamental to debt management because inflation undermines the financial measurement needed for a reasoned approach to personal, corporate and government financial planning and control. Benefits must be considered in the light of their costs.[91] British society has been lulled into a degree of economic unreality where expectations of benefits have been unrealistic in relation to the necessary *costs* by reliance on debt. Public debt differs from taxation insofar as, except in the case of irredeemable banknotes, it must be serviced by the payment of interest. This necessitates additional taxation which has powerful social effects. When tax rates are high both working and saving are discouraged and investment and economic activity are adversely affected. There is thus a strong temptation for governments to tackle the problem by increasing the growth rate of nominal Gross National Product. Congdon sees this as equivalent to debt reduction by increasing inflation.

> "... inflation eases any government's debt management problem. It reduces the real value of debts incurred in the past. This lowers the tax burden necessary to service the debt and so facilitates the task of keeping the budget deficit down."[92]

Such policies seriously disadvantage savers. In Britain while inflation "reduced the real value of central government debt by 5.3 per cent a year" between 1972-3 and 1979-80 the "real rate of return to bondholders was -7.2 per cent."[93] Supposed beneficial action in the public interest, can be very detrimental to individual interests.

But it is not just savers who suffer from inflation. Debtors may be benefited by having less to repay but only if they survive financially. The impact of inflation on financial measurement can often remove this possibility. Individual circumstances differ and movements in prices, measured by an index based upon some group of commodities, are not representative of the experience of the individual debtor. Consequently sound loan propositions can be undermined. Weston and Copeland illustrate the difference in a project evaluation resulting from a situation where revenues are subject to a 7% inflation rate, Cash costs are subject to 8% inflation and the general inflation in the economy is 6.5%. Without inflation the project would yield a positive net present value of £3642. When inflationary effects are brought into the calculation, the result is a *negative* present value of £911.[94] The message is clear. Even with modest inflation rates, unpredicted changes in government social policy, using monetary expansion to finance it can force efficient firms, with a responsible management, into serious debt and liquidation.

1.3.2. BORROWERS

The problem of debt in the eighties has been reflected in the titles chosen by contemporary writers such as *The Debt Trap* by Andrew Hartropp [95] and *The Debt Threat* by Tim Congdon.[96] These ominous titles highlight the danger of *default.*

Different Views of Borrower Default

The crisis level of the debt problem raises questions about how default arises. Attention may focus on borrowers, lenders, macroeconomics or unforeseen circumstances. Borrowers may lack responsibility, awareness or competence. Problems with impulse buying might be linked with financial recklessness where the borrower fails to engage in any financial planning. There certainly are some people who do not want to accept the limits of their own resources and want to push their financial interrelations ratio beyond acceptable limits. There may be a lack of financial awareness where borrowers fail to appreciate the true borrowing costs or fail to appreciate that they are subject to alteration with market rate rises. Regarding mortgages, the unsophisticated might assume that the security aspect covers them as well as the building society. With unsecured lending borrowers may be unaware of the personal risks involved. Borrowers may be well intentioned, but lacking in competence in financial planning or budgeting. When debt default occurs there may be an unwillingness to accept the necessary disciplines to rectify the situation.

It might be suggested that the fault is not with borrowers at all. Many bankrupts would claim that they were overtaken by circumstances outside of their control or that they were let down by their bankers. There are a variety of ways in which lenders may be at fault. Credit promotion practices might encourage irresponsible borrowing. The advance approvals mechanism might not be sufficiently thorough. The borrowing costs may be very high. In the event of payment difficulties the lender may precipitate a worse default situation by premature actions. Ways in which the macroeconomic situation may impact adversely upon the situation of debtors through interest rates and inflation have been considered in previous sections. Legislation and taxation can also have adverse effects by altering disposable income or tax reliefs. There are also problems of unexpected changes in individual circumstances resulting in personal crises affecting the ability to service debts. In 1989 alone an estimated 2.4 million adults had at least one debt problem.[97]

The current literature assessing the social impact of personal debt illustrates many of these points and will be considered later. This literature does not, in the main, address the morality of debt but summarizes the findings of empirical research, and comments upon it. A significant proportion of the literature about debt is concerned with private individuals. The material covers consumer credit, credit cards, mortgage debt and specific cases such as utilities. Major reports such as the *National Consumer Council* publication *Security risks: Personal loans secured on homes* and the *OFT* reports *Overindebtedness* and

Consumers' Use of Credit Survey provide a general picture of the situation at the time of writing. The *Institute of Business Ethics* addressed the subject in a conference report entitled *Personal Debt - Is it too much encouraged?* There are a variety of publications that deal with the impact of debt upon particular groups such as Gillian Parker's "Unemployment, low income and debt", and "Making ends meet: women, credit and debt". The *Policy Studies Institute* has engaged in a debt research programme since 1981 and now has a list of publications including:

(1) *Credit and Debt* which reviews domestic spending and consumer credit use.
(2) *Water Debt and Disconnection* on recovery policies and practices.
(3) *Gas Debt and Disconnection* which looks at reasons for being cut off.
(4) *Paying with plastic: A study of credit card debt.*
(5) *Money Advice and Debt Counselling* reviews current practice in Britain.

Such publications indicate the seriousness and the scope of the debt difficulties and the need to tackle the root causes of such a widespread problem. Particularly useful resources with authoritative empirical data or informed comment on the experience of debt are summarized below.

(1) *The Indebted Society: Credit and Default in the 1980's* (Janet Ford) Sociological orientation with a primary focus on causes and types of consumer debt problems.
(2) *Credit, Debt and Poverty* (Richard Berthoud) Survey report providing statistics and comment upon them.
(3) *Credit and Debt: The Consumer Interest* (National Consumer Council) Report addressing problems arising from the way in which credit is marketed.
(4) *What to do when someone has debt problems: A practical survival guide* (John McQueen of *The Bankruptcy Association*) A book in popular style packed with practical suggestions.[98]
(5) *Families in Debt: The Nature, Causes and Effects of Debt Problems and Policy Proposals for their Alleviation* (edited by Andrew Hartropp) and its digest, *The Debt Trap: The Impact of Credit on Families and Policy Proposals for the Future.* Research findings and comment addressing the ethics of debt from the standpoint of three Christian principles: justice, mercy and hope.[99] Includes a practical recovery plan.[100]
(6) *Aspects of Credit and Debt* (Geraint Howells, Iain Crow and Maureen Moroney) Aims to increase understanding of the problems created by the credit society. Covers choices available to poorer consumers, responsibilities of lenders, borrowers and regulators.[101]

Different Types of Borrower

Distinguishing the Family from the Corporation. The consideration of borrower responsibility requires some exploration of the differences between private households and organizations The extent to which a family can be distinguished from business corporations remains a matter of debate. Beutler, Burr, Bahr and Herrin propose seven distinguishing characteristics of the family, namely:

(a) generationality and permanence;
(b) concern with the totality of individual members;
(c) orientation that is linked to caregiving;
(d) emotionality;
(e) emphasis on qualitative process;
(f) altruism; and
(g) governance that is linked to nurturance.[102]

Similarities between the Family and the Corporation. The strong emphasis upon concern, care, and altruism would seem at first sight to be distinguishing factors but Roma S. Hanks and Marvin B. Sussman have sought to show that examples exist of behaviour parallel to each of these characteristics within a corporate context, and that corporations share with families such characteristics as altruism, emotionality, caregiving and so on.[103] It is not surprising that the individual within a particular community, whether domestic or corporate, should behave in similar ways. Human action involves certain characteristics irrespective of the context. Matthews, Goodpaster and Nash have drawn attention to the fact that the general manager of the business organization must: *A.* face outwards to the external environment of the organization; and: *B.* face inwards to the internal environment of the organization. They distinguish the ethical relevance of this as follows.

> "In *A*-transactions, the general manager views the corporation as a *moral agent* in the wider society, while in *B*-transactions, the general manager views the corporation as itself a *moral environment* to be managed with a view to the freedom and well-being of its members."

Policy formulation and policy implementation are involved in both types of transaction. The authors find in this an acknowledgement of the fact that the corporation is a microcosm of the community in which it operates in some respects and a macrocosm of the individual citizen working in that wider community in other respects.[104]

Conclusions. These considerations show that we should not drive too sharp a wedge between families, corporations and government. They are all centres of human action and involve human behaviour. Taking caring is an example, the moral expectation would be that human action in each sphere would be caring. This is reflected in the view that corporations should be concerned about the social impacts of their actions. Likewise expectations about government action include social policies. The same is true with respect to rational activities such as planning which might relate to the logistics and financing of the family holiday, to a credit promotion campaign or a new road programme. In each case we would expect a reasoned step by step approach. There are differences too. We do not expect the same level of financial competence in a family as we expect from corporate managers. The banker needs to recognize that, with respect to families, the customer side of the transaction is not organizational but domestic. This is more personal to the customer than work situations and bank actions may have much deeper impacts. In developing a general model for debt operations we will recognize both similarities and differences among borrowers.

1.3.3. LENDERS

Shareholders

The technical values of bank management are defined in terms of asset, liability and capital management. Madura explains that the managerial decisions involve a risk-return trade-off and he thus deals with an integrated balance sheet approach that covers liquidity management, interest rate risk management, the management of exposure to default risk, and bank capital management. [105] These must be translated into appropriate practices at the operational level. Our task is to consider what is necessary to ensure ethical bank operations management, especially in connection with lending. In this day to day activity macro-ethical issues of debt management are no more than a background beyond the control of the immediate participants. Lending officers can, however, influence micro issues.

We have seen that lenders could be at fault by continuing to extend credit to borrowers in difficulty, relying upon the security. Asset appreciation would give the bank a comfortable safety margin to avoid loss. But would such a serene view of the deteriorating financial situation be in the borrower's best interests? We might go on to ask whether the bank should regard itself as having the kind of responsibility to his customer which places a high priority on his good both in the short and the long term? The bank manager might reply that his responsibility is to the shareholders of the bank and consequently short-term corporate profitability is necessary because, as Will Hutton has pointed out, the market tends to undervalue future profits and prioritize shareholder value for purposes of measurement.[106] However, the manager might acknowledge that he does have a responsibility to his customer and that the course of action taken has regard to that and is to his customer's advantage. The tension remains, however, that corporate profits can militate against customer care.

Stakeholders

Mutual responsibility is important in debt management. It has been prominent in the debate about debt forgiveness. Various commentators expect the banks to accept a greater measure of responsibility for the Third World debt problem. Perhaps there is a need for financial institutions to reconsider their responsibilities so those borrowers obtain a fairer deal? Does not the lender have a responsibility to ensure that intending borrowers do not over-commit themselves? Managerial answers to such questions would depend upon the corporate culture. According to H.S. Brown et.al. the latter refers to:

(a) basic values and fundamental assumptions of the members about the corporation, its environment and human nature;
(b) organizational structures and rules arising from these; and,
(c) policies and procedures to implement goals and norms.[107]

This level of inclusiveness requires an approach to constituencies that goes beyond issues of technical efficiency. If fundamental assumptions about the organization within a natural and social environment have their place then the corporate culture requires relevant resource-related social values.

If the basic cultural stance of the organization is one of bottom-line management, then the manager will see his operational responsibility as being to meet his profit targets. This need not be interpreted as a purely financial motivation. As profit is an indicator of effective operation, the manager may see profitability as an evidence of efficiency and consequently social contribution. A.C. Samli explains that in bottom-line management it is

"implied that if the firm is managed with a specific profit goal in mind, it will yield profit and will be efficient."[108]

But such an approach may be orientated to the short term and may have an exclusive focus on shareholder returns. The constituency approach does at the very least require lenders to recognize a responsibility to their borrowers. "Constituents are those to whom business leaders feel responsible."[109] It also requires managers to achieve a match of outflows and inflows of resources. The route to this might be seen, not as goal setting, but co-ordination in consultation with the various constituencies. As we have seen, the difficulty with such an approach is in moving to an operational model that has an ethical basis and can be of value in the day to day activities of a large organization such as a bank with many branches throughout the country. Of necessity, the daily lending activity, which could give rise to over-commitment, will be remote from the infrequent consultation with all of the affected parties.

But is over-commitment such a bad thing after all? There certainly have been shifts in outlook with respect to debt from the prudent Victorian stance to the post war period, which has been characterized by budget deficits until recently. This shift from debt repayment to debt dependency has been paralleled in the eighties by individuals whose total unpaid debt (excluding mortgages) more than trebled from 1981 to 1989 [110]; so the climate of market orientated lending must not be allowed to overshadow customer care.

1.3.4. BORROWER-LENDER RELATIONSHIP

Government

The UK government is a major borrower of funds for both capital expenditure and current expenditure. The *relationship* with its lenders is largely impersonal. The mechanisms for borrowing are mainly government stocks, dated and undated, Treasury Bills, National Savings and North American loans. The lenders range through Government agencies, financial institutions, companies and individuals. Some lenders are looking for highly liquid remunerative assets. The banks hold Treasury Bills for this purpose. Some lenders are looking for secure fixed return investments. Gilts answer to this requirement and may be held by charities or investment portfolios among others. Some lenders are looking for a safe haven for small savings. Many private individuals invest in National Savings for this purpose. When we look at the *balance* of this borrower-lender relationship it can be seen that it is very much dictated from the borrower side, although this is not to say that the government can determine all of the factors unilaterally. The government decides upon its borrowing requirement and issues the appropriate instruments to cover it. We have already seen that government borrowing is embedded within the macroeconomic picture. This provides the government with considerable scope to manage its debt. Debt Management has included the use of the debt to stimulate the economy in line with Keynesian demand management. The government is also in a position to influence *interest rates* and this affects the cost to itself of the borrowing in which it engages. During the period 1931 to 1947, for example, the British government pursued a low interest policy. Debt *repayment* takes place as dated stocks mature. This does not necessarily mean that the burden of debt is reduced because new issues replace maturing securities. Thus over the five year period 1990 to 1994 sterling debt rose as follows [111]:

1990	£185,870 million
1991	£190,475 million
1992	£205,359 million
1993	£224,211 million
1994	£283,559 million.

With governments in *default*, debt repudiation is a possibility, but it involves political and economic costs. Repudiation involves breach of contract and such action undermines the social stability based upon contractual relations. Just as non-payment damages the credit of individuals and firms, government repudiation of debt destroys investor confidence and it becomes much more expensive to raise finance. We have already seen that governments do have alternatives to repudiation because of the capacity to manipulate the economy. A high inflation policy has the effect of reducing, in real terms, the amount which has to be repaid.

In assessing the *significance* of the relationship, as we noted earlier, if government attention is focused exclusively upon technical economics in a pragmatic way, the sense of obligation to balance the books and repay debt is likely to decline. The economic necessity of matching income and expenditure or assets and liabilities is not seen as something to which there should be a strong moral commitment and higher market rates then impact upon everyone. Because of the size and influence of government transactions and their influence upon interest rates and inflation otherwise sound lending propositions can result in disaster.

Business

The borrower-lender *relationship* between the banker and the business customer is one of interaction. The normal pattern of sanctioning and controlling advances will involve meetings between representatives of the bank and the management of the firm. There is thus a more interactive and personal relationship that helps to build mutual trust and confidence. This should help to support a long-term stable relationship with adequate communication. The *balance* of the relationship is more equal than that dealt with in the previous section. The banker is not in a position to dictate to the customer who can always transfer to the competition. The balance in the relationship can change, however, if the customer is in difficulties, as the possibility of moving to another bank will disappear. The *interest rates* will be subject to negotiation within a range dictated by market rates, over which neither the banker nor the customer has control. Large companies will have a stronger negotiating position than smaller firms. *Repayment* will be an important part of the ongoing relationship. The banker is lending funds that are repayable on demand or at short notice and cannot countenance too much hard core borrowing where no inroads are being made by repayments. In the event of *default*, non-repayment will lead to recovery action by the bank and damage the customer's credit rating for the future.

How *significant* is the nature of the borrower-lender relationship in this case? It is evident that the two parties have a measure of equality of standing. There will be something of a spectrum in this connection. The self-employed trader will not necessarily have the same managerial and financial competence as the directors of a medium-sized or large limited company who will be able to relate to the bank manager on a more equal footing. If there is a transparent exchange of information between the business firm and the banker there is certainly opportunity for a thorough review of the lending decision and the ongoing financial position. This should provide a basis for preventing default which is in the interests of both parties. As already indicated, the macro economic situation will also have a bearing but adequate sensitivity analysis should help to prevent over-commitment unless unpredictable changes of government policy upset the calculations. Both parties would be wise to anticipate such a possibility because the extent of present government intervention makes it difficult for the government to do anything without major economic consequences.

Individuals

The borrower-lender *relationship* with respect to personal lending is now ambiguous. It all depends upon the particular way in which the customer borrows. Credit cards and credit scoring are impersonal but where individual appraisal by interview is used there will be more interaction. The *balance* between the parties is very much on the side of the banker. He will usually have greater financial expertise and has the last say in the loan transaction. The customer still has the option of taking his business elsewhere but is unlikely to find very widely different terms by going to another bank. *Interest rates* are a major problem for the personal borrower. The risk premium built into the rates makes borrowing expensive unless security such as a legal mortgage on property is offered. Rates for credit cards and personal loans are high, typically over 15% APR with inflation as low as 2% and 3-month interbank rate around 6%, and this significantly increases the cost of associated purchases. The linkage with market rates with respect to loans has had serious ramifications for many borrowers. Increasing default on the part of personal borrowers compounds problems as banks increase lending rates in order to compensate for a higher level of bad debts. *Repayment* has various patterns for personal borrowers from highly structured repayment plans in the case of mortgages and personal loans to revolving credit on credit cards. The ability to manage the latter varies considerably among the population in general particularly where multiple debts are involved. *Default* may not be too serious if the borrower keeps on talking to the lender and is able to re-schedule payments. As already seen debt difficulties may lead to irrational behaviour under stress and problems can compound very quickly. The lack of a sound relationship between the parties would be a serious drawback under such circumstances.

The borrower-lender relationship is of considerable *significance* for personal borrowers. Personal objectives have to reflect resources. Society expects borrowers to manage their finances to repay their debts out of their income. Politicians and the media shape people's views. A government policy to increase home ownership can raise expectations among those in rented accommodation and if they are financially unsophisticated the need for advice at the time of contemplating borrowing will be considerable. If advertising brochures focus on possession rather than costs and risks the customer can be very badly served by the institutions.

Reliance on debt, mismanagement of debt and misleading advice concerning debt can all contribute to a mortgaged future with personal assets committed as security for advances and interest accumulating on debts faster than can be repaid. It may be, however, that high levels of debt are not to be taken too seriously. There has been an expectation among personal borrowers that inflation would reduce in real terms the amount to be paid. Sir Gordon Borrie comments upon it being one of the worst aspects of the high inflation of the seventies that it had the long-term effect of reinforcing the habit of borrowing to the hilt.[112] Realization of the changed macro economic conditions dawned too late. Borrowers then faced the issue whether to honour commitments or not. This is at the heart of an ethic of debt management which if it exists at all must include the necessity of honouring commitments. There must be no willful default, subtle evasion or outright repudiation by borrowers.

The above summaries show that there are similarities and differences between the situation of individuals within families and that of corporations and government. Some recognition of this is evidenced by references to "the public household" when speaking of national income and expenditure. This is not to suggest that there is the same degree of complexity in families, firms and nations, but it recognizes that there are sufficient points of similarity in financial matters to draw some parallels. From the lender's point of view, the consideration of different classes of borrower is not so much about principles of appraisal as the legal status of the borrower. For this reason, too sharp a distinction should not be drawn between individuals, and firms in a general model for lending to business and personal borrowers, although it will be clear that government borrowing could not be accommodated within such an operational model. It will be important to remember distinctions in relation to business firms and private individuals and adequately reflect this in appropriate refinements.

1.4. Conclusion

This chapter has illustrated the seriousness of over-commitment for all borrowers. It has clarified that interest can facilitate the allocation of finance in the context of borrowing, saving and intermediation but that we must not be complacent about the ethics of the financial system. The social history of Britain shows that debt has caused many continuing problems including: property seizure, heavy-handed recovery methods, debt dependency, interest charges, national debt growth, debtor relief and consequences such as illness and family problems.

1.4.1 FACTORS NOT SUITABLE FOR INCORPORATION INTO AN ETHICAL MODEL FOR DEBT MANAGEMENT OPERATIONS

Capitalism. The market economy and the price mechanism require a value framework within which to operate.[113]

The financial system. The usury debate is not the key to the financial system. More important is the kind of system necessary so that interest can be used in an unproblematic way. This would require macroeconomic ethical research.

Public debt management. Government macroeconomic policy has significant impacts on borrowers, and the possible *impacts* need to be considered in debt management but the *ethics* of such policy is a subject in its own right.

S*ound money* is relevant in so far as reasoned approaches to financial matters require quantitative data. Stable price levels benefit financial planning and control. This book will not be able to address the ethics of sound money but debt management needs to consider the effects of changing price levels.

Levels of interest rates are a significant problem for borrowers and can become a serious burden as a result of rising market rates and bad debts. Borrowing rates move with the former and are adjusted upwards to allow for risks of the latter. The ethics involved with the former must be studied at a macro level. We address the latter in an indirect way, by providing a basis for reducing default and consequently the risk premium.

Ethics and values implementation. We are concerned with *values* relevant to debt management not *processes* for their implementation within a particular organization. Participatory management and codes of ethics have been mentioned and might be chosen by particular institutions as part of an action plan for ethical improvement. This book is about the core values needed in debt management to generate cohesive action for improvement within the tripartite interest group of borrowers, lenders and advisers.

Loan officers do not have control of these matters. A subject such as commercial bank strategic management would require some consideration of the impact of the banks' total lending activities on public debt, sound money and interest rates as these are bound up with banking in the macroeconomic context. This book addresses *operational* management.

1.4.2 THE RELEVANCE OF STEWARDSHIP

We have encountered more than once the subject of stewardship although this has not been developed in any way. Stewardship has been recognized as being a key concept. Ronald H. Preston states with regard to the sub-human creation:

> "Human beings are called to exercise responsible stewardship of it. This is what Christian reflection drawing upon biblical materials suggests. We have had enough recent examples of human brashness in blundering about the universe without due care. Stewardship is harder to exercise than we had thought. *But it remains a key concept.*"(Italics mine)[114]

Preston is right in his emphasis upon the strategic role of stewardship but the problem is not just one of it being hard to exercise stewardship, as Preston suggests; it is that there has been a lack of commitment to stewardship. What we have said, for example, about the Protestant Work Ethic has shown that the complex phenomenon of the Calvinistic ethic concerning work and wealth can be crystallized in different ways. If this ethic is reduced to a model where stewardship is marginalised or reduced in scope, it is no longer faithful to the spiritual roots of the ethic. Fundamental to the ethic is the fact that the calling is not general but *specific*. It is a calling *to stewardship*. The reformers were explicit on this point. With respect to the things at our disposal, we possess them

> "... on the condition, that being content with a frugal and moderate use of them, we should take care of what shall remain. Let him who possesses a field, so partake of its yearly fruits, that he may not suffer the ground to be injured by his negligence; but let him endeavour to hand it down to posterity as he received it, or even better cultivated..."[115]

In order to the flourishing of this economy and diligence every one should regard himself as a steward of all he possesses. This emphasis upon a specific stewardship calling that embraces *all* possessions indicates a breadth of application that can only be properly addressed in structural and operational matters by an analysis of the values involved and an acceptance of the full set of values. We can understand how certain aspects of stewardship could have, as Weber found, an affinity with leading capitalistic ideas. Preston mentions how hard work and austerity fostered thrift which encouraged saving, accumulation and investment at a time when merchant capitalism needed funds.[116] He explains the readiness of capitalists to use those elements in Calvinism which were congenial to their own position [117] and we can see how, not only secularization, but a selective use of the concept, could produce, as he claims, an "unpleasant residue."[118] The battle for a comprehensive application of the concept was still being fought in the nineteenth century. E.J. Garnett explains how Edwin Freedley presented arguments: "for the systematization of life in such a way that a man should recognize" his stewardship "in *all* his dealings" and "not just that part specifically appropriated for religious or charitable purposes."[119] Such an approach would require the application of stewardship values, not only in domestic or church life, but also in business. One problem with the application of the stewardship concept within the business context is that it has been too narrowly focused upon efficiency. Preston, in discussing the Protestant Work Ethic sees its enduring relevance in this economic emphasis:

> "The merit of the Protestant Work ethic is its stress on efficiency. Waste is an offence to it. Efficiency is the economist's virtue. ... Of course there are other virtues besides efficiency. To pursue that to the exclusion of all others quickly leads to inhumanity. Nevertheless vested interests often cause us to pay too little attention to it, or to fudge the issue. The Protestant Work Ethic has a perennial relevance..." [120]

Preston, although he thinks the time has come to part with the Protestant Work Ethic, is too realistic about economics to dispense with the responsible stewardship involved in it. This element, in his view, remains relevant because efficiency cannot be disregarded. While agreeing with this insistence upon relevance, it is necessary to go *further* than economic efficiency. Our analysis of the stewardship concept in the next chapter will show that the concept involves much more and will show that its relevance to business operations is much beyond reducing waste and conserving resources, although these are important components. Even Preston ignores this broader aspect, seeing stewardship in a merely technical way, so the originality of my work will be apparent to the reader.

We can agree with Preston that the pursuit of efficiency on its own leads to inhumanity. I addressed this problem in my thesis, *The Dignity of Man and*

Business Enterprise in Reformed Thinking.[121] But we need to recognize that the stewardship concept itself, fully expounded, is not inconsistent with social values, but encourages socially responsible business practice and this will be shown as we proceed. The original theological milieu of the concept is much broader than economics and involves social concern even in business transactions. (*Note 113* refers) Garnett shows that one of the problems that has concerned Christians about political economy is the need for much clearer definition.[122] The same is true with respect to stewardship. Otherwise the moral emphasis is missed out. This was one concern of Victorian Christians.

> "They agreed that up to a point men consciously pursued their own interests, but denied that any harmony emerged naturally from the sum of these alone. Human motivation was complex and affected by the interplay of a variety of factors; self-interest could, for example, often be balanced by a sense of human collectivity."[123]

As we proceed it is not necessary that the concept of stewardship ignores social issues. The whole purpose of the research is to secure greater social benefit through more responsible lending. The task is similar, but not identical, to earlier endeavours "to restore meaning to concepts which had either been narrowed or distorted." As Garnett explains, while some who addressed the political economy were "naive exponents of 'the Protestant ethic'," others sought to "give that assumption closer definition and moral rigour."[124] This kind of activity requires much more detailed exposition and application, and this is what we intend with respect to the stewardship concept.

The latter, unlike much political economy, is free from the criticism that the blurring of the boundary between work and leisure results in redundancy. Stewardship applies to time, talents, skills and possessions and is by no means limited in its scope to remunerative work. It might be argued that wider choices resulting from more available time would *increase* the need for stewardship values.[125]

In the following chapter consideration will be given to the concept of stewardship as a concept with potential to provide a model which addresses such fundamentals as we have mentioned above. By analyzing the concept, a set of values will be developed which can be brought to bear on the subject of ethical debt management. The analysis will address the broad context of economic activity before looking more closely at the ramifications for business activity in general and borrowing and lending in particular. This analysis will provide a basis for looking in detail at lending operations to see what light is shed upon them by the stewardship values.

Notes to Chapter 1

1. 'Usury' in James Hastings (ed.) *Encyclopaedia of Religion and Ethics, Vol. XII,* pp.555-556.
2. 'Usury' in George A. Buttrick (ed.) *The Interpreter's Dictionary of the Bible.*
3. Psalm 109.11.
4. Deuteronomy 23.20.
5. Deuteronomy 23.19-20.
6. Nehemiah 5.7-11.
7. Leviticus 25.35-37.
8. Colin Brown (ed.) *New Testament Theology, Volume 2,* p.668.
9. Matthew 25.26-27 and Luke 19.22-23.
10. John F. Sleeman 'Usury and Interest' in James F. Childress and John MacQuarrie (eds.), *A New Dictionary of Christian Ethics,* pp.639-640.
11. Ronald H. Preston, *Religion and the Ambiguities of Capitalism*, p.138.
12. John F. Sleeman, 'Usury and Interest' in James F. Childress and John MacQuarrie (eds.) *op.cit.,* pp.639-640.
13. Paul Mills, *The Ban on Interest: Dead Letter or Radical Solution?* pp.3-5.
14. *Ibid.*, p.3.
15. Hugh Barty-King, *The Worst Poverty: A History of Debt and Debtors*, p.3.
16. *Ibid.,* pp. 8-9,15, 17, 40.
17. *Ibid.*, p.53.
18. *Ibid.*, p.61.
19. *Ibid.*, pp.50, 59, 67.
20. *Ibid.*, p.80.
21. *Ibid.*, p.85.
22. *Ibid.*, p.89.
23. *Ibid.*, pp.97, 144.
24. *Ibid.*, p.145.
25. Georgina Battiscombe, *Shaftesbury: The Great Reformer 1801-1885*, p.296.
26. Hugh Barty King, *op.cit.*, p.163.
27. *Ibid.*, p.171.
28. Natu Patel (ed.) *Central Statistical Office: Financial Statistics*, p.60, Office for National Statistics (c) Crown Copyright, 1996.
29. Andrew Hartropp, *The Debt Trap: The Impact of Credit on Families and Policy Proposals for the Future*, pp. 5-7.
30. 'Capitalism' in James F. Childress and John MacQuarrie (eds.), *A New Dictionary of Christian Ethics*, p.73.
31. Gerald J. Williams, *Ethics in Modern Management*, pp.104-111.
32. W. Michael Hoffman et.al. (eds.) *Emerging Global Business Ethics*, p.54.
33. Patricia Werhane, *Adam Smith and His Legacy for Modern Capitalism*, reviewed by William H. Baumer, pp.206,218, *Journal of Business Ethics,* 14: 1995,
34. John P. Noonan, *General & Special Ethics*, p.161.
35. Max Weber, *The Protestant Ethic and the Spirit of Capitalism.*

36. R.H. Tawney, *Religion and the Rise of Capitalism: A Historical Study.*
37. Ronald H. Preston, op.cit., pp.36-37.
38. Richard A. Slaughter,. *The Foresight Principle: Cultural Recovery in the 21st Century,* p.140f.
39. Hazel Henderson, 'From Environmentalism to Ecophilosophy: Retooling Cultures for the Twenty-First Century' in Michael Hoffman et al. (eds.) *The Corporation Ethics and the Environment.* p.292.
40. 'Social Ethics' in Bernhard Stoeckle (ed.) *Concise Dictionary of Christian Ethics*, p.236.
41. Amartya Sen, *On Ethics and Economics*, pp.6-7.
42. Ahmed Riahi-Belkaoui, *Morality in Accounting*, p.8.
43. Cecily Raiborn et al., *Managerial Accounting*, p.673.
44. Jon Elster (ed.), *Rational Choice,* Washington Square: New York University Press, 1986 quoted in Mary E. Guy, *Ethical Decision Making in Everyday Work Situations*, p.30.
45. Sharon H. Garrison, *The Financial Impact of Corporate Events on Corporate Stakeholders,* p.16.
46. Amartya Sen, *op.cit.*, p.12.
47. *Ibid.*, p.13.
48. Robert E. Lane, *The Market Experience,* pp.49-50.
49. David Parker, 'Chaos Theory and the Management of Change'. Unpublished paper dated May, 1994, p.17; subsequently published with the same title, pp. 1-13 *Proceedings of the Lunar Society, Vol. 8*, Sept., 1998 reproduced here by kind permission of David Parker.
50. *Ibid.*, p.18.
51. George Watson, *The English Ideology: Studies in the Language of Victorian Politics* pp.68-69.
52. Asa Briggs, *Victorian People: Some Reassessments of People, Institutions, Ideas and Events 1851-1867,* p.222.
53. Will Hutton, *The State We're In*, pp.241-242.
54. Milton Friedman, *The Counter-Revolution in Monetary Theory*, p.24.
55. Milton Friedman in Robert J. Gordon (ed.), *Milton Friedman's Monetary Framework: A Debate with His Critics*, pp.61-62.
56. F.A. Hayek, *Denationalisation of Money: The Argument Refined*, pp.117-118.
57. Ronald H. Preston, *op.cit.*, p.150.
58. Bill Richardson and Peter Curwen, 'Do Free-Market Governments Create Crisis-Ridden Societies?', pp.551-560, *Journal of Business Ethics,* 14, 1995.
59. John Atherton, "New Occasions Teach New Duties?: 9. The Individual and the Organization.", pp.356-362, *The Expository Times,* Sept., 1994.
60. Samuel C. Certo and J. Paul Peter, *Strategic Management: Concepts and Applications*, p.223.
61. William M. Evan and R. Edward Freeman 'A Stakeholder Theory of the Modern Corporation: Kantian Capitalism' in Tom L. Beauchamp and Norman E. Bowie, (eds.) *Ethical Theory and Business*, Fourth Edition, pp.81-82.

62. James W. Kuhn and Donald W. Shriver, Jr., *Beyond Success: Corporations and their Critics in the 1990s*, p.76.
63. *Ibid.* p.77.
64. *Ibid.*
65. Adrian Cadbury, 'Ethical Managers Make their Own Rules' in Andreas R. Prindl and Bimal Prodhan (eds.) *Ethical Conflicts in Finance*, p.41.
66. John Donaldson, *Business Ethics: A European Casebook*, p.v.
67. Christopher J. Cowton, 'The Development of Ethical Investment Products' in Andreas R. Prindl and Bimal Prodhan (eds.) *op.cit.*, pp.216-217.
68. Institute of Business Ethics, *Report on Business Ethics Codes,* 1996.
69. John Donaldson, *op.cit.* p.79.
70. *Ibid.*, pp.79-80.
71. Bimal Prodhan, 'Ethics, Finance and Society' in Andreas R. Prindl and Bimal Prodhan (eds.), *op.cit.* p.17.
72. Tim Congdon, *The Debt Threat: The Dangers of High Real Interest Rates for the World Economy*, pp.3-4.
73. *Ibid.*, pp.4-5.
74. *Ibid.*, p.7.
75. *Ibid.*, p.9.
76. Paul Mills, *Interest in Interest: The Old Testament Ban on Interest and its Implications for Today.*
77. Jeff Madura, *Financial Markets and Institutions*, pp.34-37.
78. Tim Congdon, *op.cit.*, p.10.
79. *Ibid.*, pp.10-11.
80. F.W. Pointon, 'Conflicts of Interest for Lending Bankers' in Andreas R. Prindl and Bimal Prodhan (eds.), *op.cit.*, p74.
81. Tim Congdon, *op.cit.*, p.19.
82. Susan George, *A Fate Worse than Debt*, pp.14-29.
83. Alan C. Shapiro, *Foundations of Multinational Financial Management*, p.618.
84. Susan George, *The Debt Boomerang: How Third World Debt Harms Us All.* p.64. In *Faith and Credit: The World Bank's Secular Empire* (San Francisco: Westview Press, 1994), Susan George and Fabrizio Sabelli argue that the policies of the World Bank are in fact exacerbating the problem.
85. Tim Congdon, *op.cit.*, p.27.
86. *Ibid.*, p.35-36.
87. John M. Keynes, *General Theory of Employment, Interest and Money.*
88. Tim Congdon, *op.cit.*, p.45.
89. *Ibid.*
90. Ronald H. Preston, *op.cit.*, p.117.
91. Tim Congdon, *op.cit.* p.50..
92. *Ibid.*, p.64.
93. *Ibid.*, p.67.
94. J.Fred Weston and Thomas E. Copeland, *Managerial Finance*, pp.132-133.
95. Andrew Hartropp, *op.cit.*

96. Tim Congdon, *op.cit.*
97. Credit Action *Newsletter*, undated.
98. See Bibliography for details of the above items.
99. Andrew Hartropp (ed.) *Families in Debt: The Nature, Causes and Effects of Debt Problems and Policy Proposals for Their Alleviation*, pp.7-12.
100. *Ibid.*, pp.129-141.
101. Geraint Howells, et al., *Aspects of Credit and Debt.*
102. I.F. Beutler et. al. 'The family realm: Theoretical contributions for understanding its consequences', pp. 805-816, *Journal of Marriage and the Family,* 51(3), 1989, in Roma S. Hanks and Marvin B. Sussman, (eds.), *Corporations, Businesses and Families*, p.5.
103. Roma S. Hanks and Marvin B. Sussman, 'Where Does Family End and Corporation Begin: The Consequences of Rapid Transformation' in Roma S. Hanks and Marvin B. Sussman, (eds.), *op. cit.*, pp.6-7.
104. John B. Matthews, Kenneth E. Goodpaster and Laura L. Nash, *Policies and Persons: A Casebook in Business Ethics*, p.619.
105. Jeff Madura, *op.cit.*, pp.475-498
106. Will Hutton, *op.cit.,* p.161.
107. Halina S. Brown, et al. *Corporate Environmentalism in a Global Economy*, p.152.
108. A. Coskun Samli, *Social Responsibility in Marketing: A Proactive and Profitable Marketing Management Strategy*, p.47.
109. Verne Henderson, *What's Ethical in Business?*, p.52.
110. Credit Action, *Mind Over Money: Participant's Notes*, Session Four, p.1.
111. Office for National Statistics, *Annual Abstract of Statistics*, p.315, (c) Crown copyright 1996
112. Gordon Borrie, 'Law and Morality in the Marketplace' in Andreas R. Prindl and Bimal Prodhan (eds.) *op.cit.*, p.46.
113. Moltmann goes so far as to call Calvinism the religion of *socialism* and maintains that Weber understated the Puritan emphasis upon "responsibility for the community, care for the weak, and education for the common good..."See the discussion in Michael Novak, *The Spirit of Democratic Capitalism,* pp. 260-262. It is certainly unwise in the light of work such as that of Paul Seaver, 'The Puritan Work Ethic Revisited', *Journal of British Studies*, pp. 35-53, 19(2) to *assume* that the puritan theology was favourable to a motivation based upon self-acquisitive accumulative economics. Of the merchant community Seaver speaks as follows. "For all that they may have experienced occasional bouts of anxiety over particular incidents of sharp practice, they found, on the whole, comfort in a ministry that praised their responsible stewardship and their generosity toward their poor neighbors." *ibid.* Seaver's article reminds us that we must be careful not to make versions filtered by the hearer, into the whole message proclaimed.

    That the Puritan Calvinism did have a strong communal emphasis including mutual concern and support is clear from the primary materials. The samples below illustrate the point, which could be supported by many pages of such quotations from different sources. The ones below are taken from Matthew Henry, *An Exposition of the Old and New Testament.* This commentary originally published at the beginning of the eighteenth century came to be used in many puritan homes and continues in use now. "Even the light of nature teaches us to be civil and respectful towards all, though they be strangers and sojourners. The noble generosity of these Canaanites shames and condemns the closeness, and selfishness, and ill-humour, of many that call themselves Israelites." *Volume I, Genesis 23. 3-15* "If men borrow large sums to trade with, to increase their stocks, or to purchase land, there is no reason why the lender should not share with the borrower in his profit; or if to spend upon their lusts, or repair what they have so spent, why should they not pay for their extravagances? But if the poor borrow to maintain their families, and we be able to help them, it is certain we ought either to lend freely what they have occasion for, or (if they be not likely to repay it) to give freely something towards it." *Volume II, Nehemiah 5.1ff.* "He is a *just man,* who has not, by fraud and under colour of law and right, *oppressed any,* and who has not with force and arms *spoiled any by violence,* not spoiled them of their goods or estates, much less of their liberties and lives, v7. ... Nay, he is one that has not lent his money *upon usury,* nor *taken increase* (v.8.), though, being done by contract, it may seem free from injustice ... yet, as far as it is forbidden by the law, he dares not do it. A moderate usury they were allowed to receive from strangers, but not from their brethren. A just man will not take advantage of his neighbour's necessity to make a prey of him, nor indulge himself in ease and idleness to live upon the sweat and toil of others, and therefore will not take increase from those who cannot make increase of what he lends them..." *Volume IV, Ezekiel 18.1ff.* These quotations remind us that we must be cautious about taking the elements selected by subsequent authors for purposes of analysis as being the whole story. There was much more to the puritan social ethic than correspondences with individualistic economic capitalism. It included a strong sense of what would today be called *social responsibility.*

114. Ronald H. Preston, *The Future of Christian Ethics,* p.94.
115. John Calvin, *Calvin's Commentaries: 1. The Pentateuch,* p.22.
116. Ronald H. Preston, *op. cit.,* p.119.
117. *Ibid.,* p120.
118. *Ibid.,* p134.
119. E.J. Garnett, *Aspects of the Relationship between Protestant Ethics and Economic Activity in Mid-Victorian England,* p.221.
120. Ronald H. Preston, *op. cit.,* p.134.
121. W.R. Mohon, *The Dignity of Man and Business Enterprise in Reformed Thinking.*
122. E.J. Garnett, *op.cit.*, p.12
123. *Ibid.,* p.15
124. *Ibid.,* p.268.
125. See Ronald H. Preston, *op. cit.,* p.133

CHAPTER 2

THE STEWARDSHIP CONCEPT

2.1 Definitions of Stewardship in Business

2.1.1. STEWARDSHIP ACCOUNTING

Analysis of the stewardship concept as found in the accountant's world locates our preliminary analysis *in* the practitioner's world. The two main branches of accounting are financial accounting and (cost and) management accounting.[1] The former addresses the past, whereas the latter addresses the present and future. Financial accounting reports on what has happened, whereas management accounting provides information to determine actions for success. Management accounting provides information to managers for planning, decision and control, whereas financial accounting provides information for investors and other stakeholders including lenders and creditors. There is overlap. Performance appraisal by the management accountant uses final accounts prepared by the financial accountant. Both branches have a planning and decision aspect but at different levels. Management accounting's management information system enables managerial planning and decision, whereas financial accounting reports help shareholders make decisions about future investment. Approaches to the function and concepts of accounting thus require some flexibility.

Richard Mattessich [2] identifies the stewardship programme as an important research tradition in accounting. This tradition sees the function of accounting as the principal's monitoring of managerial stewardship. The managers, as agents, act on behalf of the investors, controlling resources to secure a return on their investment. Accounting records transactions to provide financial information about the managers' integrity and diligence in the use of business assets by periodic reporting on this stewardship. Mattessich discusses the tradition under three phases: the plain periodization approach, the original agency approach and the agency-information approach. [3]

The plain periodization approach presupposes that monitoring managerial stewardship requires costs and revenues to be appropriately allocated to periods. It consequently concentrates on historical cost, periodical allocation of revenue, costs, consistency, objectivity and so on. It is made operational through matching costs and revenues within a given period using historical values. The accountant is not viewed as a valuer but one who presents the financial facts of the period in coherent financial statements with an appropriate level of objectivity. K.H.M. Naser states that:

> "*Stewardship accounting* came into existence to guarantee both the physical safety of assets and prove that those who administrated them had done their job properly."[4]

Implicit in this statement are the ideas that corporate managers administer assets on behalf of shareholders, that the relationship sustained is one of stewardship and that this relationship involves accountability. The essence of the stewardship involved is accountable administration of resources in the interest of others. The asset base must be preserved, its use is expected to enhance its value and an account of how things stand is required periodically. There is thus a close relationship between *investment*, stewardship and accountability. Shareholders invest in the company expecting the management to exercise due diligence in its stewardship of investor funds. Management must invest these in income-producing assets and administer them wisely to enhance the value of the firm for the benefit of the ownership group. An account is given of the outcome by providing relevant information.

That *information* is accumulated in the financial accounting system. The importance of financial accounting as a check on managerial stewardship received impetus from the 1844 *Joint Stock Companies Act* that led to a divorce between ownership and control in connection with incorporated business enterprises. The investor ownership group, who contribute the capital, do not engage directly in the resource deployment which is performed by managers. There is thus a need to check on managers to ensure that their use of funds does not deplete, but rather enhances the shareholders' net asset value. This is particularly important, as resource deployment cannot be divorced from the financing of the assets. Stewardship is not about asset utilization *in isolation from* the firm's sources of funds, whether share capital or borrowing. A particular project can only be profitable if its returns exceed the costs, including loan capital costs. Managerial stewardship is thus intimately interwoven with debt management.

Accounting statements are used to monitor managerial performance. Key documents are:

(1) The *Balance Sheet* showing the period end financial position to check sources of finance, deployment of funds and asset inventory.

(2) The *Trading and Profit and Loss Account* showing the periodic annual performance that reveals the managerial decision taking quality.

(3) The *Cash Flow Statement* that provides a check on cash movements reveals any diversion of liquid funds.

These statements interest stakeholders including investors, bankers and creditors. For investors they provide financial information to assess managerial performance. Changes in the relative importance of these statements illustrate the complexity in accounting for stewardship. The balance sheet attracts greater interest in times of financial uncertainty when short-term liquidity problems may threaten the company's immediate survival or long-term solvency, impacting on the safety of shareholder funds. The trading and profit and loss account provides information about the income and assumes greater importance as a guide to assessing the shareholder return on investment and expected dividends. Elliott and Elliott identify profit as the key measurement of stewardship.

> "Managers are the stewards appointed by shareholders. Income, in the sense of net income or net profit, is the crystallization of their accountability." [5]

There is here a broadening of the idea of accounting for stewardship. Traditional stewardship has, according to Michael Bromwich, three main functions, requiring management to demonstrate that:

(a) The enterprise has accounted for its resources in the correct way.
(b) The enterprise has utilized resources only for legitimate purposes.
(c) The enterprise has correctly maintained its assets. [6]

In this *legal* view, emphasis is on compliance with legislation and accountancy standards in presenting financial information; and compliance with the objects of the company in the guardianship and use of the asset base. The quotation from Elliott and Elliott illustrates, however, that investors are also concerned about net income, which provides funds for dividend distributions and increases net asset value. For them managerial stewardship involves asset guardianship, but also opportunity seeking and competent decision taking in asset utilization. Earnings per share have thus become one of the cardinal investor statistics. An interpretation of the stewardship role, by the *American Institute of Certified Public Accountants* focused upon the *economic* aspect stating that: "Stewardship refers to the efficient administration of resources and the execution of plans for conserving and consuming them." Chen suggests that this could be viewed as "a higher form of stewardship - an accounting for effective use not merely custody or maintenance,"[7] and Lehman adds,

> "Such statements contrast with the aim of maintenance or safeguarding of these resources, implying that stewardship and the reporting of stewardship should be redefined as the effective utilization of a firm's resources."[8]

Managerial stewardship consequently requires planned strategic action and a broader view of the stewardship objectives of the accounting system emphasizes the monitoring of *performance*. This requires income reports that facilitate the distribution of income generated by operations, recognizing that earnings are not only a shareholder objective but a determinant of shareholder behaviour and investment strategy. Shareholders cannot be marginalised as stakeholders without threatening the survival of the firm by losing their funding. The relationship between shareholders and management is thus fundamental.

2.1.2. AGENCY THEORY

Agency theory looks at the stewardship relationship between managers and shareholders in terms of principal and agent. The principal is the party to the contract on behalf of whom, and in whose interest, the agent acts. The shareholder group is thus viewed as principal and the management as agent. Mattessich discusses recent developments under two stages: his second stewardship programme phase, namely the original agency approach and his third stewardship programme phase, the agency-information approach; the periodization approach, already considered above, being his first. Mattessich describes the original agency approach as predominantly descriptive, which, when combined with information economics, yielded the agency-information approach which focuses upon the contractual relationship between principal and agent and the risk sharing between them. For our purposes these two phases are sufficiently similar to be discussed together. The distinction is primarily one of the extent of the analysis, the original agency theory being the more descriptive of the two. The accounting-information system has, according to Mattessich, a vital role in the original agency approach because the latter is preoccupied with the costs arising from the different goals of principal and agent and the desire to produce contracts that minimize conflicts by motivating managers to actions beneficial to shareholders, thus reducing monitoring costs.[9]

The agency-information approach is Mattessich's third phase of the stewardship paradigm. It combines the enhanced statistical decision theory of information economics with the original agency approach to analyze the contractual relationship between principal and agent seeking to establish an efficient contractual basis.[10] Mattessich draws attention to the following relevant factors.

(a) The problem of moral hazard. As complete monitoring of management by shareholders is seldom possible, the former may maximize personal benefit to the detriment of the principal.

(b) The problem of management having an information advantage in terms of access and understanding may lead to selective communication of financial data to the principal resulting in contractual disadvantage.

(c) The recognition that the capital risk might be spread more widely than the shareholders (the principal) and include lenders. [11]

Mattessich's discussion of the second and third phases of the stewardship programme underlines the relationship of trust between shareholders and management, and shows the need to ensure that there is no breach of that trust. The latter requires monitoring of the contractual relationship between principal and agent so stewardship accounting's major focus is on the need to hold management accountable and reduce breakdowns of trust.

> "Audit actions were no doubt devised in the belief that the scrutiny of a person's report of the discharge of his responsibilities was beneficial both for him and for the interests he served".[12]

This review of the stewardship concept in the accountant's world has introduced us to three perspectives. There is the *legal* perspective with its emphasis upon *compliance*, the *economic* perspective with its emphasis upon *resource utilization* and the *agency* perspective with its emphasis upon *relationships*. These different yet associated perspectives suggest that it is possible to think in terms of a *moral* perspective and analyzing the stewardship concept to determine relevant business values. This is implied in Lehman's book *Accounting's Changing Role in Social Conflict*. The need for a coherent set of resource-related social values capable of application to business operations is met by the stewardship concept, ethically considered, which has both a *relationship* and a *resource-related* aspect. The legal and agency interpretations emphasize the former, with respect to the owner-manager relationship; and the economic interpretation emphasizes the latter, in connection with the resource management activity, *but these definitions remain at the technical level*. They suggest a role for the stewardship concept in determining business values, and encourage the expectation that an analysis of the concept's moral implications will provide a coherent set of values to apply to business operations for purposes of *ethical* analysis. We have discovered.

(1) The core idea of stewardship for the accountant is the *accountable* control of resources mindful of owner interests. The financial accountant must record and process financial facts to make accountability transparent and present a true and fair view of the firm's position the shareholders.

(2) Stewardship accounting points to the e*ntrustedness* in the relationship between shareholders and management. The managers' position of trust requires a sub-ordination of self-interest to the shareholders' property interests. Managers have no absolute control of assets.

(3) The legal view of stewardship is still prominent in accounting practice and determines many financial accounting information system requirements. The key idea is compliance but with the underlying values of *integrity* and

faithfulness in the performance of one's duties as demonstrated by accurate accounting, compliance with company objects, and asset base guardianship.

(4) The economic perspective forbids a limited a view of stewardship. Managerial stewardship involves the use of assets in a way that will enhance their worth. Shareholder interest thus requires *effectiveness* of action in the control of assets. Beyond custody are opportunity seeking, strategy, planning, decision and organization to secure development reflected in the *enhancement* of asset value.

(5) Agency theory illustrates stewardship in terms of the principal-agent accountability relationship. It focuses on *trust* by drawing attention to the potential for a conflict of interest and points to the importance of *fairness* in the reliability and availability of information so that shareholders are not disadvantaged in the contract with management. The Agency theory, in recognizing the interests of parties beyond the shareholders, shows the need to broaden our consideration of fairness and accountability beyond the ownership group to make more explicit *social* responsibility.

(6) The statements of financial position and financial performance, necessary for stewardship accounting, must give a true and fair view of the corporate affairs and this underlines the importance of relevant and accurate measurement if correct judgements are to be made.

We conclude that, thinking ethically about the concept of stewardship provides us with values relevant to business activity which can accommodate factors relevant to both resource and social perspectives. It helps us to see what is necessary in human relationships and in dealing with resources and a set of values begins to emerge including accountability, entrustedness, integrity, faithfulness, effectiveness, enhancement, social responsibility and reliable measurement.

2.2. A Critical Defence of Stewardship

We first defend our stewardship approach from the criticism that it has no originality in comparison with other views and then make the case that it can be defended in the light of debates about the failure of stewardship, the need for social responsibility and the problem of common values.

2.2.1. STEWARDSHIP DISTINGUISHED FROM OTHER VIEWS

It might be argued that our stewardship approach is tarnished with the same problems as reductionism in economics, that it is impractical social altruism or nothing more than financial prudence or technical stewardship. We address each of these below to show that, although we are not foisting an alien concept onto commercial activities, we are making more explicit the concept's *moral* significance to apply it to operations. The various values come into play in each particular situation around the central role of measurement.

Reductionism in Economics

We argued in *Chapter 1* that stewardship is not to be confused with instrumental rationalism. Stewardship does not involve making exchanges in the marketplace the determinants of values.[13] It does not have to be construed as involving cut-throat competition but is just as compatible with pursuing mutual benefit through cooperation.

Impractical Social Altruism

This is not to say that we are ignoring the importance of measurement and the usefulness of the marketplace. We are not marginalising measurement and quantification in ethical decision taking. The stewardship values are applied along with quantification and vice versa to inform the decision, addressing both resource *and* social considerations. The necessity of valuation in this context requires some market values, if there is to be any objectivity of measurement. Resource constraints must be brought into our moral decision-making. Our stewardship links to service but in so doing accepts the practicalities of economic reality.

Financial Prudence

Our stewardship is more than financial prudence that emphasizes caution and protection from risk. As we saw above, stewardship must aim at development, and financial prudence is but one factor of the overall ethical decision-taking. There are risks of action but stewardship requires action and aims at the most beneficial actions. It is necessary to address risk but stewardship accounting is not restricted to measurement to eliminate risk but includes measurement to assess net benefit. To secure relevant benefits it may be necessary to borrow, but the matching of benefits and disbenefits is a key aspect of the overall decision. This, as we will show, involves more than financial prudence.

Technical Stewardship

Stewardship as a moral concept must also be distinguished from our starting point in stewardship accounting. In the latter the emphasis is legal and economic and necessarily materialistic. It has to do with accounting for resources and distribution of financial gains between management and shareholders. This materialistic focus is but one aspect of stewardship which, when viewed from an ethical standpoint, introduces a much broader range of responsibilities. The stewardship concept was found compatible with market economics as a price mechanism providing *financial* values but not with the market as deciding *moral* values. In the business context operational stewardship requires matching but we are not reducing stewardship to technical finance but, the strength of our position is that: (a) we are recognizing the necessity of stewardship matching for economic measurement and giving it an important guiding role, while (b) insisting upon a broader view of stewardship. Our analysis is not a form of reductionism that lets figures rule, but is moral in expecting the figures to serve as a guide as we pursue ethical decisions. Although they must retain a central role in operations we maintain that they are there *because* we want to bring to bear entrustedness, effectiveness, enhancement, accountability and fairness.

2.2.2. STEWARDSHIP DEFENDED FROM CURRENT CRITICISMS

We now consider the contemporary scene and dismiss related criticisms of our stewardship approach addressing failures of managerial stewardship, the need for corporate social responsibility and the problem of common values. With respect to each of these debates there is a case to be made for stewardship values. The first two are closely linked. Stewardship is prominent in the contemporary policy agenda because of problems in business relationships, particularly regarding accountability. There are two main concerns. The business community has witnessed some massive defaults in managerial stewardship and there is the concern that, if management focuses too exclusively upon responsibilities to shareholders, other values will be overlooked to the detriment of customers, employees and society.

Managerial Stewardship

Criticisms of managerial stewardship must be taken seriously but it is necessary to understand that the problem arises from the limitations of technical stewardship and that this does not militate against our moral approach, but underlines the need for it. We can only mention some headline cases.

Guinness (1985-1986). Senior executives entered into arrangements with investors to purchase shares in Guinness and unlawfully used the company's funds to support its share price.[14]

Blue Arrow (1987). Senior NatWest executives were criticized for misleading advertisements and failure to disclose relevant information. [15]

Barlow Clowes (1988). Two Barlow Clowes funds collapsed resulting in almost £200 million of investor funds being placed at risk. [16]

BCCI (1991). The Bank of Credit and Commercial International was closed down in sixty-two countries by regulators. Price Waterhouse auditors concluded that management had been deceitful and that accounts were being used to fraudulently route funds. [17]

Such cases illustrate that contractual relationship and legal constraint do not, of themselves, answer to the need. The two opposing sides in the debate about more legislation or greater voluntary self-regulation can accept the importance of entrustedness as a value. Tom Cannon states that directors are expected to act:

> "in loyalty and good faith in performing duties
> with skill and care in managing assets
> and that as trustees
> they cannot profit from their position of trust." [18]

Regarding their relationship to shareholders, the law upholds the latter's contractual rights and the accountant provides financial information for monitoring and investment decisions. We, however, do not see stewardship in a restricted legal way. The criticism that stewardship only addresses managerial responsibility to *shareholders* is not applicable to our broader ethical view of stewardship entrustedness, the essence of which is that the directors will have regard to the interests of all providers of funds and be accountable for the funds and any increase on them. This is not exclusive to relationships legally defined in terms of principal and agent but enshrines the more general sense of entrustedness and the accountability associated with it. Debenture holders, for example, are not part of the ownership group but they have entrusted funds to management as part of a loan contract. This is not a principal-agent contract but a debtor-creditor relationship but the latter involves both entrustedness and accountability and it would be unethical for managers to ignore claims of these providers of funds because they are not part of the ownership group. Our view of stewardship allows for creditor interests, including director accountability for the safety of the funds and any agreed percentage increase on them.

We can go further with our broader ethical stewardship and say that it could handle other cases of entrusted funds. So, for example, in the case of BCCI, the management failure involved a systematic neglect of the responsibility to *depositors* who had entrusted their savings to the bank for safekeeping. The further question arises of responsibility to other stakeholders who may not have invested, loaned or deposited funds but have some interest in the operation of the business in some way. This broader management responsibility is usually addressed under the general heading of social responsibility and we want to show next that our stewardship values are not incompatible with corporate social responsibility, but promote it.

Stewardship and Corporate Social Responsibility

The concept of social responsibility is not new to British managers.[19] Issues such as pollution, consumer relations and training have been on the agenda for some time.[20] Managers are familiar with such issues and will have reflected upon moral values relevant to the firm's objectives in these areas. Stewardship might be criticized as hostile to social responsibility and a narrow view of entrustedness gives this impression. Milton Friedman, for example, refers to social responsibility as "a fundamentally subversive doctrine"[21]; but it still expresses the realization that firms do not operate in a vacuum. Brenner and Molander surveyed *Harvard Business Review* readers finding that 83% agreed that social responsibility is an issue that concerns the role the corporation should play in society.[22] Andrews defines its scope in the following way,

> "The words mean in part voluntary restraint of profit maximization. More positively, they mean sensitivity to the social costs of economic activity and to the opportunity to focus corporate power on objectives that are possible but sometimes less economically attractive than socially desirable." [23]

Andrews includes: voluntary contributions to charity, ethical operation beyond legal requirements, the inclusion of imputed social worth in decisions about economic opportunity and quality of life in the corporation.[24] The criticism is heard that managerial stewardship would undermine social responsibility but this need not be the case. Andrew's explanation shows that the term "social responsibility" can be used to cover a wide range of matters and we must show that regarding our stewardship view, managers' goals can be broader than the narrowest view. Theodore Levitt, in an article entitled, 'The Dangers of Social Responsibility' argues for limited business goals.[25] For him the function of business is "to produce sustained high-level profits" without which business cannot survive.[26] He argues that narrow ambitions are acceptable.

> "The trouble is not that it is too narrowly profit-oriented, but that it is not narrowly profit-orientated *enough*."[27]

Levitt favours every possible step to prevent business becoming involved in welfare and society instead of long-run profit maximization The coalescing of business and government is seen by him as an open door to the domination of society by a professional corporate bureaucracy and something to be avoided.[28] Three issues here are relevant to our view of managerial entrustedness, namely: profit orientation, profit maximization and social constraints on the latter. The business firm must be *profit-oriented*. Only by matching costs and benefits can management engage in planned stewardship. Profit here identifies business activities that represent efficient allocation of resources. Even non-profit-making organizations use cost-benefit analysis to furnish financial justifications of projects using surplus net present value. Profit, here is a measure of efficiency to help eliminate waste and secure the maximum output from limited resources. This suggests that *profit maximization* should be considered as beneficial in relation to the use of resources and our stewardship model could approve this general statement. Is this the same as saying that profit should be pursued to the exclusion of social considerations? We do not have to concede this, and, in saying so, we do not have to part company with all business thinking, because it is not necessarily the way in which business practitioners have seen themselves according to the analysis of Tom L Beauchamp and Norman E. Bowie in a chapter entitled 'Ethical Theory and Business Practice'. They argue that business practitioners *have seen themselves* as making a significant social contribution through their activities. These authors recognize that the moral principles embedded in public morality are usually present in a "vague and underanalyzed form".[29] We cannot expect to find the moral philosopher's precision in the general ideas that have surfaced in business thinking when businessmen seek to provide ethical justification for pursuing their personal benefit. Ethical theories do not work through into practice with the precision that philosophers would like but contribute to an amalgam of ideas drawn from other disciplines as well. One way in which business has sought to provide a moral justification for its activities is to combine self-interest with community benefit through what they term utilitarianism. This illustrates a disposition favourable to our broad ethical stewardship.

Adam Smith put forward the view that competing individual interests would promote the public good. Beauchamp and Bowie summarize his position as follows:

> "... egoism in commercial transactions leads not to the war of all against all, but rather to a utilitarian outcome, that is, the largest number of benefits for the largest number of persons."[30]

Notwithstanding the rejection of Smith's optimism by many business managers, Beauchamp and Bowie found that a basic kind of utilitarianism has registered in the business practitioner's consciousness. This is not surprising. Although utilitarian concern with pleasure (Hedonistic utilitarianism), happiness (Eudaemonist utilitarianism), or knowledge, friendship and courage (Pluralistic utilitarianism) are more remote from business, *utility*, understood "in terms of the satisfaction of individual preferences", is not.[31] In the mind of business executives the latter is closely associated with maximization through efficiency. These two authors do not overlook the fact that utilitarianism is concerned at a deeper level with more than efficiency, which is simply an instrumental good, but explain the relationship, as seen by some business executives, as follows.

> "The means to maximization is efficiency, a goal that persons in business find congenial, because it is highly prized throughout the economic sector. Efficiency is a means to higher profits and lower prices, and the struggle to be maximally profitable seeks to obtain maximum production from limited economic resources. The utilitarian commitment to the principle of optimal productivity through efficiency is an essential part of the traditional business conception of society and a standard part of business practice."[32]

In this way a form of utilitarian maximization of the good could find a close affinity with business practice through the technical value of efficiency which is regarded as a means to maximization. It could be argued that the utilitarian commitment referred to has no morality left in it, and that the main correspondence lies in the use of the same term. This kind of depletion is part of the problem of business ethics. Terms with *moral* content are reduced to *technical* values. As we argued in *Chapter 1*, even Preston limits the value of stewardship to technical efficiency,[33] but we have shown that the concept is, for ethics, much more significant than this. The obvious advantage of efficiency has resulted in its being a business goal and this goal has the appearance to business practitioners of being consistent with the good of society as a whole and one justification of competitive business activity. Several matters need to be distinguished in connection with this explanation. We might have expected a technical principle of optimal productivity through efficiency to register with business executives as a result of the study of microeconomics. It would not be surprising, therefore, that in so far as business practitioners formed any conception of society they would do so on the basis of a utilitarian conception of society. It is interesting to note that in an earlier edition of their work, Beauchamp and Bowie had inserted after the second sentence of the above quotation the explanation: "These produced goods and services are presumably intended to promote the greatest good of the greatest number."[34] It is important to notice the words *presumably intended to promote*, in connection with *the greatest good of the greatest number*. These two writers obviously recognize that a greater quantity may not promote the public good. The *nature* of what is being produced also has an important bearing upon the value of the activity. What is to the point in Beauchamp and Bowie's analysis is that, in so far as business practitioners have sought a moral justification, they have sought it in terms of a basic form of utilitarianism. To accept that this kind of substratum exists is not to justify it. The writers are aware of the problem that utilitarianism based upon subjective preferences would raise the whole question of the formulation of an acceptable range of preferences because maximum output from scarce resources is only one factor and the nature of the outputs is also fundamental to the community good. However, the idea that egoistic behaviour can yield utilitarian results, in their sense of community good, appeals to at least some business practitioners, in their search for a moral justification, ostensible or real. It provides for a vigorous pursuit of self-interest in competition with others in the confidence that this will lead to maximal utilitarian results. We are not attempting to justify such a view of business morality but would argue that it indicates some readiness to face up to such issues as our broader view of stewardship would raise.

The influence of utilitarianism upon economics doubtless has a bearing upon why business executives would find a basic utilitarian framework congenial as a context for justifying self-interested competitive activity. The reality, however, is that there are many influences that run counter to maximization in practice. We mentioned in our discussion of *Agency Theory* that managers may prioritize personal benefit above that of the firm and shareholders. In owner-managed firms, which account for 50% of UK GDP, retaining management control ranks above maximizing income. Research has shown that managers are often content to accept a satisfactory level of achievement in the place of profit maximization (*Satisficing*) or prioritize divisional performance above company performance (*Suboptimization*). Research in the US showed that 80% of companies have more than four goals. Even where profitability is a primary goal it is not necessarily *profit* maximization but improved efficiency that is sought.[35] We may conclude that many managers already accept goals beyond profit. We do not have to agree with the above attempted synthesis of egoism and utilitarianism to recognize that it involves a place for what society wants. The brief sketch of how business has attempted to justify the morality of enterprise reveals that such a view accepts *some* responsibility to society on the part of business. Our argument is not that there is, or will be, a correspondence between maximized outputs and the social interest. It is that, in so far as business seeks any justification in terms of the public good, it affirms a responsibility for social contribution beyond self-interest. The vigorous pursuit of profit is seen as benefiting society at large because it will lead to more efficient production, which will raise output levels, so that more goods and services will be available from the limited resources used. As long as the producer is responsive to market demand, which reflects individual preferences, the utility of individual consumers will be maximized. Not only will the quantity of goods and services be increased, but, *according to this view*, the goods and services will match what people want, and living standards will be improved in accordance with their aspirations. In this way, it would be argued, enterprise would serve the public interest, although what people *want* is not necessarily the same as the social *interest*. It is unlikely that the management of lending institutions would reject some such commitment to the social good; indeed, it is more likely that they would claim to accept a responsibility to serve the social interest by helping people to satisfy their legitimate aspirations. There is a wide spectrum of opinion as to which aspirations are legitimate and attention will be given to those relevant to debt management in later chapters. What concerns us is the need for bank managers to recognize stewardship values beyond efficiency, measured by profit, so that they see that efficient service provision is not the only value bearing upon social responsibility.

This brings us to the crux of the matter. We are not denying the importance of the measurement by matching which gives rise to a profit figure but our stewardship model places that within a set of values and this involves mutual interaction in decision-making. Surprisingly, support for this can be found in Milton Friedman, an advocate of the narrower view of business responsibility.[36] For him, in a free economy,

> "there is one and only one social responsibility of business - to use its resources and engage in activities designed to increase its profits so long as it stays within the rules of the game, which is to say, engages in open and free competition, without deception or fraud."[37]

Friedman sees anything wider than this as undermining the foundations of a free society. He argues that business managers are not in a position to know what the social interest is and are not civil servants.[38]

Andrews definition of social responsibility has shown that there is much of it that is not about social *projects* and the key phrase for socially responsible *stewardship* is Friedman's reference to the "rules of the game"[39] which implies a good deal more in connection with social responsibility than "open and free competition without deception or fraud"[40] might suggest. In the relationship between an institutional lender and a private borrower the former is possessed of considerable financial expertise, whereas the latter is normally very limited in financial knowledge. The traditional role of the banker has consequently included financial advice with the position of trust that this implies. In this context social responsibility is not as clear cut as avoiding cartels and dishonest dealings. It would appear that U.K. bankers have recognized this. They have accepted responsibility for improving public understanding of their role through the *Banking Education Service*. Debt problems focus attention on a broader concept of the "rules of the game", where there are two parties of unequal standing involved. It is legitimate to expect bankers, with their financial expertise, to feel a sense of responsibility to society at large. Debt difficulties bring family hardship. Traditionally bankers have not put profits before customers' interests, although loan sharks profit by doing so. The rules of the banking game have a breadth that might be described as a social responsibility "other than making maximum profits for stockholders" [41]. Banks have traditionally conducted their business this way. Managers can be found who will have persuaded customers not to borrow, not to save their banks a bad debt, but to save their customers from a level of over-commitment which would have yielded the bank its return but would have been very painful for the borrower. The real issue, therefore, is not whether business takes itself to have a social responsibility, but the scope of the responsibility, and, with respect to lending, we must go beyond law, economics and technical efficiency. This will be done in later chapters.

Stewardship within a Common Morality

We now address the criticism that our stewardship model would not be of interest to practitioners. The importance of values in business is emphasized in the Business Ethics literature. According to John Donaldson values drive business.[42] There are two immediate problems. The first is that we might not approve of the values driving business. Profit maximization might be advanced as the firm's mission, but if this is interpreted to mean the pursuit of maximum profit irrespective of who is disadvantaged in the process, this will attract many critics. The second problem is that as soon as there is agreement that broader values are needed, there is difficulty in reaching consensus as to which values come into this category. Michael Novak has underlined this problem by drawing attention to the danger of imposing upon business alien values, which are remote from an ethic for business which is relevant to the nature and aims of business enterprise as a specific kind of activity.[43] Such a core of values in an organization or industry is not peripheral because voluntarily adopted shared values are desirable to produce cohesion of action. This was illustrated above by Agency Theory in relation to the stewardship research tradition in accounting. If the principal must exert total monitoring in order to secure an agent activity consistent with the principal's interest, then this will involve significant costs. As long as dysfunctional behaviour persists the organization will be at a competitive disadvantage in relation to firms that have eliminated it by cohesive action. Organizations thus have a vested interest in seeking core values internally and within their respective industries. The acceptance of this latter point is illustrated from the "level playing field" approach which is often introduced to argue for common stances by firms within an industry so that all are operating on the same footing. It is encapsulated in the *Stock Exchange* motto "My Word is My Bond" which reflected the common commitment to fidelity in connection with oral bargains struck in the market. The economic community cannot function if it leaves individuals to make their own rules of conduct.

The source of the particular values in business presents problems for business ethicists. John Donaldson goes so far as to say that our dominant forms of business "seem to rest on incompatible foundations" and that "the particular values chosen ... can be thought of as arriving from a variety of sources."[44] We can certainly see something of this in connection with Ethical Investment. The investment portfolios of financial institutions offering ethical funds reflect the investment managers' assessment of what will be *operationally* attractive. The value profiles of ethical investment portfolios reflect the interests of particular pressure groups or matters of public concern such as conservation, animal rights, anti-armaments, and so on. Investors of different ethical standpoints opt into particular funds and it is not necessary for investment managers to convince their clients to opt into a comprehensive underlying ethical viewpoint in order to secure their commitment to a particular set of values. Stating this fact is not to suggest that we should be content as a society to follow pressure group activity in adopting values for business operation, but the ethical investment situation alerts us to the way in which acceptance of an operational set of values does not require the portfolio managers to convince the investors to accept some underlying ethical system. Managers and investors take sets of values according to how they fit in with their own operational requirements if the values commend themselves to their moral consciousness. There is advantage and danger here. The advantage is that people of different ethical standpoints can find common ground. The danger is that practical business ethics is reduced to a "pick and mix" morality without any central core of agreement. What we have been showing is that stewardship provides such a core set of values.

MacIntyre is an example of a philosopher who very typically has sought to address the need for values that are not just individual. He argues for "a mandatory cultural quest for values and virtues that will rescue us from individualistic, relativistic moral theory."[45] In the search for relevant values MacIntyre directs us to history. At an individual level we give explanations of our actions by the stories that we tell concerning our reasons for them. Kuhn and Shriver explain how we are able to reflect upon what is good, right and valuable in our individual lives by seeing actions in their context.[46] In the context of such stories the reader can appreciate the significance of the value issues that arise in relation to the different parts of the advancing narrative.

Although it is not always the case, they explain how human stories from different centuries, contexts or individuals can disclose a "kinship" that enables the values that arise in their stories to illuminate our lives as we perceive how such values transcend the particulars of our immediate situation and provide an undergirding to it. This is relevant to business institutions which are supported by the surrounding culture.

> "Institutions such as the market 'float', so to speak, in *culture,* in a tossing sea of beliefs, values, meanings, and stories that therefore provide a common environment for the diverse, vital elements of the lives of persons, activities of institutions, and movement of whole civilizations."[47]

Though in a democratic society there are many different stories, told from different perspectives, diversity does not exclude that overlap which results in relevant values being able to secure common acceptance. This can be demonstrated with respect to values associated with stewardship.

We saw in *Chapter 1* that Ronald H. Preston accepts the continuing relevance of stewardship. Regarding the Protestant work ethic he states:

> "... at bottom it is concerned with the efficient use (or stewardship) of relatively scarce economic resources compared with the possible alternative uses of them. This is a perennial problem in any society."[48]

The matter of economic efficiency cannot be sidestepped. Stewardship is not, however, the preserve of economists. As we shall see in the following section, it is also familiar to lawyers concerned with the legalities of corporate governance. In relation to environmentalism, there have been newer voices too, with a more moral note. In 1990 President George Bush commented in *A New Era of Environmental Stewardship*:

> "Concern for environmental protection and stewardship has grown, solidified, and deepened over the last two decades. ...The 1990s must be the decade when we focus our attention on finding the most cost-effective, most efficient ways to prevent pollution, to reduce risks to human health and the environment, and to achieve environmentally sound, sustainable economic growth."[49]

Thomas F. Malone in *Stewardship: Paradigm for the Nth Millennium* viewed stewardship as *the* concept for consideration and reexamination.[50] Links between the environment and debt are explored in detail by Morris Miller in *Debt and the Environment: Converging Crisis.* He finds the prospects for environmental stewardship unpromising:

> "When the 'costs' of environmental damage are subtle and in any case felt only in the longer-term, and when the benefits of environmentally-damaging activities are financially tangible and immediate, it is not difficult to fathom which option the stressed-out debtors will choose."[51]

Yet, the crisis itself underlines the importance of the kind of stewardship that strikes a balance between human and environmental needs, individual and social needs, present and future needs and family and financial needs. Peter Selby, in his book *Grace and Mortgage: The Language of Faith and the Debt of the World*, explains that corrupt governments have contributed to debt crises by their poor stewardship of borrowed funds.

> "Their love of prestige projects, whatever the relevance or irrelevance of these projects to the actual needs of their people, their resort to high levels of military expenditure not least in order to ensure their own survival, and their practice of financial self-aggrandisement, have all been part of what has saddled their countries with burdens of repayment which they have no prospect of being able to bear."[52]

If the savings of some are to help meet the needs of others, if potential earnings are to provide for present asset acquisition and, if unacceptable social costs of debt are to be avoided, the implications of stewardship for debt management must be addressed and in addressing them we can be sure that we are starting with a concept and talking in a language that can be very widely understood.

2.3. Development of the Concept of Stewardship

We have looked at criticisms of stewardship and found them inadequate with respect to our ethical stewardship approach. We are now going to develop our stewardship model by looking at product stewardship, religious stewardship and environmental stewardship. We have seen that the stewardship concept has given rise to an important accounting research tradition and that it features in the corporate governance debate. In a recent publication, *Stewardship: Choosing Service over Self-Interest*, Peter Block uses stewardship to develop a new model of governance, and advocates making ourselves accountable for service to the larger organization in partnership with others independently of control or compliance. He sees stewardship as a "set of principles and practices which have the potential to make dramatic changes in our governance system." [53] In this section mention has been made of ethical investment and *Friends Provident* has used a stewardship focus for its ethical portfolio which is promoted under the name *Stewardship Fund.* Business practitioners are now also familiar with product stewardship. These by no means exhaust the examples of the use of the concept as a focus in connection with ethical activity. Outside of the business context examples include religious stewardship and environmental stewardship. These examples show that stewardship values can attract a wide and varied body of opinion and that ethical analysis of the concept and the application of the resulting set of values need not be construed as an approach that would have very limited appeal.

We will consider briefly product stewardship and religious stewardship and then provide a more extended treatment of environmental stewardship to exhibit the core values.

2.3.1. PRODUCT STEWARDSHIP

Product Stewardship is very different from stewardship accounting. It is a proactive risk management system that aims to enhance the customer-supplier relationship by taking responsibility for all impacts of the firm's activities:

from design
through manufacture
and use
to disposal

It is well established in the chemical industry and the Chemical Manufacturers Association has their own *Product Stewardship Code of Management Practices 1992.* The concept emphasizes responsibility for health, safety and the environment. Responsibility *to whom* cannot be defined in terms of any single stakeholder because product stewardship spans the total life of the product and the total range of impacts during manufacture, use and disposal. Whereas the period of use may principally affect the customer it may also affect the local community. Impacts during manufacture and disposal may principally affect the community. Stewardship in this context is used to emphasize responsibility for safety, the environment and conservation and the willingness to account for the practices and performance of the firm.

2.3.2. RELIGIOUS STEWARDSHIP

John Passmore in his book *Man's Responsibility for Nature : Ecological Problems and Western Traditions* finds little evidence that Christianity emphasizes moral responsibility with respect to handling the lower order of creation. He goes so far as to say,

> "It is certainly a mistake, indeed, to describe as 'typically Christian' the view that man's duty is to preserve the face of the earth in 'beauty, usefulness and fruitfulness'."[54]

The concept is, however, applied to economics in one contemporary North American Christian tradition that has drawn on Cornelius Van Til. Van Til made a good deal of use of the *vicegerency* concept. We will briefly outline a few of the points that he makes. Van Til maintains that humans are, on a creaturely scale, in the image of their Maker with intellect, affections and volition.[55] Our relationship is one of dependence but also of likeness or resemblance.[56] In the matter of personal action we are not drifting aimlessly on an ocean of chance but are rational beings living before a purposeful supreme Being. This according to Van Til, is what should shape our ethical ideal of self-realization, in particular in the world of business. As Van Til explains,

> "Just as the manager of a growing business needs to increase with his business in alertness, stability and comprehensiveness of decision, so man, with the development of his progress toward his ideal, would have to develop momentum in order to meet his ever increasing responsibility."[57]

The nature of this responsibility is seen when we address the subject of vicegerency. Van Til makes a point of relating dominion to vicegerency and consequently to stewardship.[58] Man is faced with "setting and keeping his powers in motion in order to reach that goal that has been set for him..."[59] Van Til sees vicegerency as the most important aspect of this program.[60] Vicegerency has the following properties in common with stewardship: *relationship*, *control* and *purpose*. The basic idea of vicegerency is that of delegated authority. It involves a relationship between parties where authority is committed to the vicegerent to exercise control in a specified sphere with a view to particular ends. Stewardship modelled upon such vicegerency should not, therefore, sink into a willful exploitation of the world. We were never intended to have absolute sovereignty. The relationship is one of delegation. The kingdom is not ours. This has implications for the nature of the control. Dominion is not to be exercised in a destructive and polluting fashion but in a constructive way. The relationship is also relevant to the goal which is in view and we are to follow ends beyond self-gratification.[61] With respect to the stewardship of natural resources vicegerency lays emphasis upon a responsible and conserving control, blending human action and the natural environment. It points to ends beyond self-interest, recognizing that life is not about getting something for nothing, by ignoring the parameters within which we are to operate, but about service.[62]

There is an application of this kind of stewardship by Harold L. Johnson in an article in the *Harvard Business Review* entitled 'Can the Businessman Apply Christianity?' Johnson maintains that Christianity has relevance for the way in which stewardship is applied. We made reference above to the importance of the nature of a relationship both for control and purpose. When stewardship is considered in a Christian context the vicegerency relationship imposes limits on both control and on our goals. It limits our control of natural resources by denying an absolute government to humans. In terms of position, though we have real control, we are *serving* in another's kingdom. Johnson thus sees stewardship as reinforcing trusteeship. We are not acting on our own behalf. The concept also acts as a limiting concept in connection with our purpose by preventing the absolutising of lesser goals.[63] The acknowledgement of a sovereign creator involves denying absolute ownership to humans. This has implications for priorities and possessions. With respect to priorities, Johnson sees our stewardship role as warning us against the idolatry of putting first in life such things as the business firm, the nation, free enterprise or one's career. With regard to possessions, Man is but the "temporal owner" of resources, a "trustee or steward". As such he is responsible to a higher authority.[64] This we can see would also have implications for the use of resources in relation to the human community. It would surely be incompatible with the Divine purpose for mankind in general to conceive of ourselves as acceptable stewards while controlling resources in a purely self interested way?

This section has shown the possibility of deriving values from the stewardship concept and this is a key point for my work. We have seen that exploration of the concept from a moral perspective can furnish values relevant to business practitioners. Even in this brief survey a variety of values have begun to be identified such as: responsible deputizing, preservation, personal development, constructive activity, conservation, unselfishness and trusteeship. It is my contention in this book that the stewardship concept with a moral emphasis, such as is found in the Christian religion, can furnish an operational framework for business managers even if not personally committed to Christianity. The relevance, value and originality of the book is that it presents a stewardship model that can build in the ethical dimension throughout, in contrast to the highly technical approach of Friedman and others.

2.3.3. STEWARDING AND THE ENVIRONMENT

One of the ways in which people think about stewardship is in connection with the environment. Stewardship is part of a broad picture involving the whole of our environment. We will show that considering stewardship of the environment can advance our argument by helping to show the key values and their interactions in decision-making. Contemporary debate has focused upon the importance of responsible asset conservation accompanying the utilization of natural resources for purposes of wealth creation. This debate is ethical as well as economic.

It is important to have the correct view of the stewardship concept as applied to ecology. Christopher D. Stone has emphasized the lack of legal rights of environmental objects and has argued that they lack the criteria of rights-holders.

> "They have no standing in their own right; their unique damages do not count in determining outcome; and they are not the beneficiaries of awards."[65]

The fact that Stone should find it necessary to speak in this way about legal rights for natural objects suggests that he feels that the natural objects are in need of protection from humans. Dominion is frequently manifested in an exploitative way and, consequently, there is discomfort with the concept of dominion and rightly so. Once severed from its root in man's vicegerency it leads down a path of self-assertive possession where ethical stewardship is forgotten. The blame for this has in fact been laid at the door of Christian ethics and Lynn White argues for a rejection of man's dominion if the ecological crisis is not to worsen.[66] He argues as follows. Christianity inherited a story of creation from Judaism. This included the idea that the physical creation serves human purposes. According to White, Christianity established a dualism between man and nature, insisting that man is permitted to exploit nature for proper ends. By the destruction of pagan animism the way was opened for nature to be exploited "in a mood of indifference to the feelings of natural objects." White sees the only route back as being the rejection of the Christian axiom that "nature has no reason for existence save to serve man."[67]

White's model is too simplistic. Although the Judeo-Christian tradition did influence science and technology, and although the application of the latter has led to environmental degradation, there is no direct link between the doctrine of dominion and the latter.[68] A.R. Peacocke has argued the case against White [69] and makes the following points: (1) The exploitation of the natural environment with its consequent adverse ecological changes has occurred from primitive times and is not specifically associated with Judeo-Christian societies. (2) There has not been a uniform attitude to science and technology among the followers of the Judeo-Christian tradition. Peacocke concludes that the Judeo-Christian tradition does not, in fact, depict man's dominion as simply brutally exploitative.[70] As Peacocke states,

> "Although 'dominion' has ... kingly reference, it is a caring 'dominion' exercised under the authority of the creator..." It is thus associated with the ideas of vicegerency, stewardship and trusteeship.[71]

We may, however, concede that White's article prompts us to reflect upon the true nature of dominion.[72] We must, however, not only get the concept right but must act upon it. One Christian writer puts it this way,

> "... it is true, as Lynn White points out, that much 'Christianity' is worse off in the area of ecology than animism. The animists think there are spirits in the trees and so they do not cut down the trees carelessly. So far as ecology is concerned, we must admit that he is right: much 'Christianity' has treated nature with less restraint than animism, not because Christianity does not have an answer but because we have not *acted on* the answer..."[73]

We must conclude that there is a way of thinking of stewardship and the environment which is exploitative and there is a way that is caring. The approach will be determined by the associated values and too narrow an approach can be damaging. This brings us to the heart of the matter.

2.3.4. THE STEWARDSHIP CONCEPT AND ITS CORE VALUES

Ethical Stewardship

We have moved from a merely accounting, legal or economic approach, through product, religious and environmental stewardship, to ethical stewardship. We now want to see by reference to the preceding, how we should define stewardship. Norman L. Geisler, in his book on Christian ethics states,

> "Our 'dominion' is one of faithful stewardship of natural resources. We are over but also in the natural world and are called upon to protect and preserve it. The biblical command to control it does not mean to corrupt it. Our power over nature does not confer the right to pollute it. On the contrary, the Christian has the responsibility to care for and keep the natural world."[74]

Vicegerency is a position of "entrustedness". We have no absolute sovereignty but are temporal owners of resources as trustees or stewards. We can thus define "entrustedness" as accountable control of resources and fill out what is involved. In looking at the subject of environmental stewardship the resources involved are those of the natural world. We must use them to survive. We can control them although even with modern technology such control remains for the moment limited. The writers we referred to recognize this dominion. What is at issue is whether this dominion results in destruction or enhancement of the material world. The concept we have outlined places a constraint upon the exercise of dominion by its emphasis upon entrustedness. This rules out absolute domination. By reinforcing a sense of trustee responsibility it encourages an approach which has more in view than self-gratification. It broadens horizons beyond immediate satisfactions and suggests an approach to nature that will embody a more caring and constructive use.

Trusteeship carries with it at least the idea of keeping intact. In human affairs the trustee has a duty to exercise the utmost diligence in managing the trust property involving its protection, preservation and distribution. The idea of keeping intact has been prominent in connection with thinking about the natural environment and has given rise to the conservationist and preservationist movements. The latter have been concerned to save the natural environment from the intrusion of humans so that their activities will not destroy the natural habitat. They would have all left, as far as possible, in its primitive condition. Conservationists, by contrast, do not oppose the use of natural resources for economic purposes but seek to prevent unnecessary damage to the environment and unnecessary wastage. Environmental trusteeship could certainly sympathize with both views in connection with the emphasis upon maintaining the resources entrusted to our care. But there are qualifications necessary. We have already mentioned that humans are incapable of not using natural resources because life and livelihood depend upon them. We cannot, therefore, view environmental trusteeship as equivalent to a security guard who preserves intact by forbidding access. Preserving the environment as an untouched wilderness cannot be our only or leading consideration. In a humanly originated trust the trustees have a duty to accomplish the objects of the trust, and the maintenance of the trust funds is in order to such attainment. The trustees' task is thoughtful management of the entrusted property to ensure that the purposes of the trust can be fulfilled. Protection and preservation is to ensure availability of funds when needed. We must adopt some such view of our environmental trusteeship. We are not mere guardians but must conserve as we use to minimize waste and maximize resources to meet further calls upon them. It is but a short step from this to realize that environmental trusteeship necessitates development. A trustee in bankruptcy, for example, has the very limited role of disposing of the bankrupt's assets and collecting any disposable income to pay a dividend to the creditors. But arrangements may be made with the bankrupt so that the latter can continue in business in the name in which he was made bankrupt and split the profits in some way with the trustee. This course of action could benefit creditors. This shows the value of development. Resources in use can increase the value of the firm. The absence of development in the case of environmental trusteeship would reduce the human race to primitive conditions.

Having made clearer the meaning of the concept of stewardship, it is necessary to specify its core meaning and the differences arising in the different contexts. The key idea, common to the various applications discussed, is that of *accountable control of resources*. This implies that there is a stewardship activity when *A* acts as one to whom resources have been *entrusted* by *B* and where it would be a breach of that trust were *A* not to have regard to *the interest of B*. In order to determine whether or not *A* had respected the interests of *B* some accountability mechanism involving measurement would be necessary. Precisely what the interest of *B* might be varies with the situation as we shall now see, in the light of our findings in this chapter.

The Concept in Different Contexts

Stewardship Accounting. In Stewardship Accounting the function of accounting is regarded as being the accountability mechanism to measure whether or not

A's (the management) control of the resources entrusted to them by *B* (the shareholders) has been in the interests of the latter. In this case the interests of *B* involve asset protection and an acceptable return on the assets. The financial reports confirm or otherwise *A*'s integrity, diligence and effectiveness. The managerial stewardship may be more complex. *B* may include more than one class of finance provider. In addition to equity shareholders (*B*) there may be preference shareholders (*B2*) and debenture holders (*B3*). In the case of *B2* the responsibility of *A* is to retain asset value sufficient to repay preference capital at nominal value, in the event of liquidation, and to make profits sufficient to pay the fixed rate preference dividend at the specified percentage. In the case of *B3* the legal relationship is one of debt and the responsibility of *A* is to repay the nominal value of the debenture on the due date and to pay the annual interest at the specified percentage per annum. As managers function they must, therefore, regard the interests of the different classes that have a stake in the resources used. To address these different interests requires appropriate planning and control. Project appraisal for purposes of remunerative asset stewardship requires that shareholder capital and loan capital are both included in cost of capital calculations to determine whether or not a return can be made.

Agency Theory. Agency Theory concentrates on the contractual relationship between shareholders and managers as principal and agent. In this case *A* (the agent) acts on behalf of *B* (the principal) using the resources of the latter, entrusted to *A*, with a view to maximizing the return to *B*.(75) This theory recognizes the problem of moral hazard. *A* may not be trustworthy and may seek to maximize personal benefit to the detriment *B*. The theory shows the importance of an accountability mechanism to monitor managerial effectiveness within the contractual relationship. Failures of managerial stewardship have focused attention upon the responsibilities of directors. This has involved a closer analysis of what is expected of *A* (the top management) in relation to *B* (the shareholders). Legitimate expectations include in this context: loyalty, good faith, competent asset management, and not seeking to profit from the position of trust. As we have seen in the previous paragraph, there may be a *B2* and a *B3* and these expectations also apply to *A*'s responsibility to these different providers of funds even though *B3* is not part of the *ownership* group. This shows the importance of core values. If *B* must exert total monitoring of *A*, significant costs will be incurred to prevent dysfunctional behaviour and this will result in competitive disadvantage in relation to firms where acceptance of stewardship values results in *B's* expectations of *A* being fulfilled. A reliable steward does not require such close monitoring and we have referred above to Peter Block's development of a new governance model based upon stewardship not self-interest.

Product Stewardship. Product Stewardship is an interesting example of stewardship applied at the operational level. There is a strong emphasis upon health and safety and environmental protection. As this is a management system it is implied that the managers assume the *A* role. It is by no means easy to specify *B* in this particular example. This is because the focus is upon the product as it passes through the different phases of its design, manufacture, use and disposal. If health and safety were to be reviewed at the manufacturing stage, management would be thinking about their responsibility to the firm's employees. If the review related to the use of the product, management would have in mind their responsibility for impacts upon the customer. In connection with disposal it would be environmental impacts that would need to be assessed. Taking all of these together *B* might be specified as: the class of all persons who either use or are affected by the product at any stage of its existence. In product stewardship it is responsibility and accountability that predominate, along with having regard to the interests of others, and as product stewardship is concerned with risk management, liability would best serve as an indicator of responsibility to whom. This would link factory safety to employees and product safety to customers.

Religious Stewardship. Religious stewardship has a primary focus upon morality rather than contractual relationships, liability and accountability mechanisms as in the preceding examples. Vicegerency provides one example of thinking about human responsibility in relation to a purposeful Supreme Being. This basic idea of delegated authority involves a relationship in which *A* (the vicegerent) is to exercise control in a specified sphere (the world) with a view to the realization of ends specified by *B* (the divine being). *A* is not an absolute ruler, nor is the kingdom his. His headship is to be directed to higher ends than self-gratification. Resource management is thus service. The nature of the control and the goals in view are to have regard to the higher authority to whom the kingdom belongs and to whom an account must ultimately be given. This moral trusteeship rules out a prioritization of self-interest and an exploitative approach to the natural environment. Action consistent with such stewardship would be expected to be caring, constructive and unselfish with both a preserving and developmental emphasis.

Environmental Stewardship. Environmental stewardship usefully illustrates the problems that can arise in specifying *B* and the interests of *B*. Even without reference to Christianity, a kind of environmental stewardship can be specified.

If, for purposes of illustration, we identify *A* as any organization, private or public, that interacts with the environment with a view to development we are then presented with the problem of defining *B*. It is immediately apparent that the natural environment transcends the immediate generation and present impacts stretch into the future. Should *B* relate to the present generation? Or should *B* relate to future generations? Or would we be better to think in terms of *B1* and *B2*? We have already seen above under *Agency Theory* that there are situations where *B* has to be recognized as being complex. If we retain in the basic definition of stewardship the idea that *A* controls resources *entrusted by B* it would close the door to defining *B* in terms of future generations. Only the present generation could be regarded as giving to organizations some delegated control over resources for their benefit. But when we introduce the interests of *B* we can see that this could then become a complex *B1* and *B2* because the existing generation (*B1*) may wish the interests of its posterity (*B2*) to be taken into consideration with respect to both conservation and asset formation. The former is preferable to our legacy to posterity being a wilderness and the latter has enabled standards of living to be improved. This is not just a question of material standards of living but of the whole quality of life. Advances in material standards of living provide opportunities for intellectual and cultural pursuits that might not otherwise be possible. But there are no easy answers to what is in the best interests of *B* considered as simple or complex.

2.3.5. KEY ELEMENTS OF STEWARDSHIP FOR DEBT MANAGEMENT

We are now in a position to summarize our findings concerning (1) the general requirements for an ethical debt management model, (2) the relevant core values of the stewardship concept that can address such fundamentals and (3) how they relate to debt management.

General Requirements

The analysis in *Chapter 1* identified factors fundamental to an ethical model for borrowing and lending.

(1) The need to recognize the problems associated with debt dependence and over-commitment.
(2) The acceptance of the existing financial system including the payment of interest without being complacent about its problems.
(3) The acceptance of economic reality but not an uncritical acceptance of an autonomous price mechanism.
(4) The recognition of social responsibility.
(5) Addressing the balance between financial gain and social care by relevant

values.

(6) Taking an economizing approach to resource management without displacing social and environmental values.

(7) The importance of relationships with a focus upon trust.

Core Values

In this chapter the concept of stewardship has been analyzed to provide a model that addresses such fundamentals. We will see in later chapters that it can be helpful to borrowers and lenders. Our findings are as follows:

(1) *Entrustedness* is the key to understanding stewardship morally. Regarding relationships the focus of the model is thus upon trust. (7. above)

(2) The core idea of entrustedness is "accountable control of resources" mindful of the claims of investors and lenders. A reasoned approach to this accountable control of resources involves *Matching,* which in financially quantifiable cases involves the matching of technical accounting. The model thus addresses economic reality and can do so in a market context. (3. and 2. above)

(3) Relationships of entrustedness, understood as accountable control of resources, require *Effectiveness* in controlling resources, subordinating self-interest to the property interests of those providing the resources. (6. above)

(4) Stewardship encourages a conserving use of scarce resources to secure planned enhancement from *Development*: outputs should exceed inputs. (6. above)

(5) *Accountability* is an integral part of stewardship entrustedness and this goes beyond the legal accountability of agent to principal to the moral accountability involved in all actions. This accountability may be viewed theologically, socially and organizationally.(4., 5. and 6.)

(6) The importance of *Fairness* in all relationships is indispensable as is obvious in stewardship accounting, agency theory, and product stewardship. Compliance with accounting requirements and the necessity for financial statements to give a true and fair view enshrines underlying values such as integrity, honesty and fairness. Fairness was also found to be relevant to the availability and reliability of information. This is not limited to shareholders but recognizes the interests of other parties (4. and 6. above)

The stewardship model can thus handle relationships by *entrustedness*, *accountability* and *fairness* and resources by *matching*, *effectiveness* of planning and control and *enhancement* or development goals.

Relevance to Debt Management

We will now show the relevance of these core values to debt management. Going back to the idea of *entrustedness*, we can see that the borrower has been entrusted with the loaned funds. The lender has made available the funds and has given the borrower control over them for a specified purpose. The funds are not given as an outright gift. The lender expects the borrower to repay the funds and the agreed interest charges. The lender thus expects that the borrowers will manage their financial affairs in an efficient manner so that repayments can be made when due. As with stewardship in environmental and business matters, so here, operational efficiency will require *matching*, including matching costs and benefits and assets and liabilities. Income will have to be matched against outgoings including interest charges and repayments. The obligation to repay does not provide funds to repay and more is needed than willingness to repay. To be faithful to the trust placed in them borrowers must be operationally *effective* and this involves engaging in planning and decision, using, within an ethical framework, the quantitative matching to which we have just referred. Effective servicing of the loan will necessitate other outgoings being reduced if interest charges increase so that total outgoings do not exceed income, but this has ethical impacts in connection with family welfare. The dynamic mutual interaction of financial matching and ethical considerations is evident here and more is said about this below. In debt management *enhancement* or development cannot be narrowly viewed as increasing asset value. As we shall see in the next chapter, the development might not be quantifiable, such as is the case with total family welfare. As long as debts are outstanding to the lenders, they are right in insisting upon the *accountability* of the borrower who ought not to disregard the element of entrustedness that enters into the arrangement. We would want to add that treatment should still be fair and that *fairness* should run through all of the relationships and procedures.

We will now address the dynamics of stewardship decision-making in relation to the environment, markets, and business to demonstrate the necessity and possibility of integrating technical operational matching with the ethical values.

2.4. The Dynamics of Stewardship Decision-making

2.4.1. THE ENVIRONMENTAL CONTEXT

Complexity

Situations of developmental activity using environmental resources disclose the dynamics of stewardship decision-making. Choices for development, rather than exploitative wastefulness, are complex, as the case of hydro-electric projects show. Suppose we consider the harnessing of mountain torrents in order to generate electricity. Consumers, hospitals and others would appreciate cheaper electricity. Primitivism would object to draping the mountain with water pipes. The conservationist would approve of the use of renewable resources. A compromise suggestion might be to embed the water pipes invisibly in the mountainside. Securing this environmental benefit involves increased construction costs. At this point the ethicist's world meets the accountant's world. The extra costs will have to be quantified and their impact on electricity prices to consumers and hospitals estimated. In the case of the latter we might ask how many operations the extra electricity prices would have funded and raise a question of ethical priorities about health treatment verses scenery. This shows that the economic matching is *integral to* and *interacts with* values when we have to make ethical decisions. Not all decision-making models can handle this complexity effectively although our stewardship model can.

Differentiating

An important model for seeking to handle social and environmental complexity is Cost benefit Analysis (CBA).[76] Applied to environmental matters, this model presupposes that the project manager's view of entrustedness should be broader than profit maximization and embrace social and habitat considerations. The technique assists decision-takers by matching detractions from a project's objectives (costs) with contributions to the objectives (benefits), in order to determine whether the overall outcome is negative or positive. In order to do this all of the costs and benefits have to be quantified in some way. Michael Common, in his book, *Environmental and Resource Economics*, gives a worked example of a hydro-electric project, including construction costs, shut-down costs, maintenance costs, savings in coal, avoidance of premature deaths and loss of wilderness recreation facility.[77] In such a case quantitative values have to be imputed to qualitative costs and benefits. Passmore, having affirmed that every action costs something, approves of such a method.

> "What can properly be demanded is not that men should cease to act in ways which involve ecological risks... but rather that they should take more account of the ecological costs and benefits of their actions. And this means, once again, that they will have to make fuller use of rational Western-type methods, cost-benefit analyses or decision procedures."[78]

This should not be pursued by CBA imputing financial values to social and environmental costs and thereby *merging* moral values into technical values in order to make a decision. Our stewardship model recognizes the importance of quantitative analysis in matching resource costs and the financial values of outputs but requires that this information inform the total decision model, not take it over. Our model differentiates between the necessary technical matching based upon market prices, which gives a reasonably accurate guide for many commodities, and values that are above financial quantification.

Integrating

What we have said above is not to approve of parallel worlds, as though accountancy information and the ethical manager will never meet. The stewardship concept is not about distancing the two, but integrating them. It involves an entrustedness which requires the use of measurement. We can agree to some extent with Passmore when, concerning CBA, he concludes:

> "... to try to construct such an analysis clears one's mind, drawing attention to costs and benefits one might otherwise have overlooked. And when a form of action is limited in character and narrow in its effects, its costs and benefits can be calculated with a relatively high degree of precision."[79]

The CBA model is not, however, sufficiently robust for ethical decision taking because of its dependence upon a technical methodology. It gives some guidance for choices between *similar* projects but it cannot handle the complex dynamic interaction between ethical judgements and financial costs. The only way forward is to integrate the two in a single model that can handle both. As we will show in the following chapters, our stewardship model can do this.

2.4.2. THE MARKET CONTEXT

The Stewardship model can function within the market economy at an operational level because it takes measurement seriously. As a general concept stewardship can be high-sounding but if it is not applied in a context where there is objective measurement it is never put to the test. The matching of costs and benefits must be based upon reasonably reliable information in order to facilitate good stewardship. For a range of costs and benefits objectivity is attainable in consequence of the existence of market prices. These market values provide a basis for arriving at the estimated costs and benefits of a project and if they were not available the matching exercise would be impractical. We might illustrate the benefit of this objective quantification by reference to health services and suggest that if health service internal markets had been approached in the context of stewardship they could have been seen in a different light and operated in a different way. All participants could have recognized the principle of scarcity, the need for efficient utilization, that this necessitates matching inputs and outputs and that the accountant's and the consultant's world must meet. The former must not be free to prioritize investment in hospital car parks over patient beds, but the latter cannot be free to make ethical decisions about treatment independently of the resource implications. The stewardship model bridges into both worlds without commercializing life. It uses prices and financial measurement to inform ethical decision but not to define values. Economics has a valid place but it is only one dimension and must not be absolutised. The market place should not determine our goals but its price information has validity within the total decision making process. We are not looking to the market place to make our moral decisions for us but to provide the price information which is necessary to engage in planning and decision which will be conserving and promote efficient operation.[80] Within a stewardship context the internal market would not be the sole criterion but would provide valid quantitative information about where more resources were needed and where they were being used efficiently. Markets can facilitate efficient resource allocation, and, by providing market values in terms of equilibrium prices, provide data which enables stewardship planning, decision and control to take place and provide a basis for accountability.

2.4.3. THE BUSINESS CONTEXT

We turn from the environment and markets to consider the ways in which what we have seen with environmental stewardship is applied in business. We have seen that measurable *entrustedness* is the essence of stewardship which involves: (a) resources being made available in a context of *matching* costs and benefits, quantified in relation to the market, and control over those resources being granted, (b) *effective* use of the resources in the attainment of relevant *enhancement*, (c) *accountability* for the use of the resources and (d) maintaining *fairness* throughout.

What we find when we turn to business enterprise is a technical approach to these values, centered on financial matching. The ownership group commits their funds to management. The managers are in a position of trust. Resources have been committed to their charge, in the understanding that, by responsible control of the resources they will not only conserve the firm's assets, but also increase its net worth. The public limited company invites savers and institutional managers to invest through its prospectus. Its relationship with the outside world is defined in its memorandum of association and the *objects clause* defines the purposes of the company. It is on the understanding that the management will pursue the specified objects that investors entrust their funds to them. In many cases this no longer places any significant constraints upon the management because the objects clauses are widely drawn. The expectations of the investors that the firm encourages through its prospectus are consequently largely in terms of *financial growth*. The significant contents, in the case of an existing business, would be, the profits and losses for each of the preceding five years, the assets and liabilities in the last set of accounts, the dividends on each class of share over the previous five years and the forecast profits. This is not to say that constraints in connection with social responsibility and ecological policies will not be of interest to investors but it would be a remarkable departure from reality to suggest that the overriding expectation of investors will be anything other than that growth which will furnish them with a return on their investment. In the final analysis managers know that they are expected to manage their companies efficiently and thus secure growth and must therefore possess suitable tools for operational decision taking. The financial growth imperative results in the precedence of the quantitative basis for decision.

Managers may be motivated to act in sub-optimal ways preferring shorter term less profitable projects to longer term more profitable projects because such projects enable the managers to demonstrate their own competence in a shorter time.[81] In addition the management of the firm may be efficient or inefficient and consequently successful or unsuccessful. In order to determine which is the case, measurement must take place. The audit function has developed in connection with company accounting to ensure that the financial reporting to the investors by the management, in order to demonstrate that they have acted in accord with the trust reposed in them, is accurate. By accounting for the various movements in the assets and liabilities of the firm the efficiency and reliability of the management is monitored and the ownership group can be assured that their wealth is being used in a responsible and beneficial way. Thus, in connection with the presentation of the periodic financial reports, the auditing accountants have a responsibility to ensure that the financial statements present a " true and fair view" of the financial affairs of the firm. If they do not, then the statements will be of little value in measuring the stewardship effectiveness of the firm's management. As with cost-benefit analysis the financial reporting statements match costs and benefits; but, in the case of business operation, costs are limited to expenditures and there are no imputed values for social and environmental costs. Ethical values thus often have a kind of parallel existence within the firm sometimes being be brought in as a filter *at the end* of the decision process.

This is far from our expectations of managers in the light of what we have said about stewardship. Our stewardship model intends that every decision will be an ethical decision, in which neither financial matching nor relevant social costs and benefits are marginalised. We want managers to accept our broad view of entrustedness beyond responsibility to shareholders so that, in their control of resources, they will take into consideration the moral dimension, and moral accountability in decision-taking, even though their financial data ignores external social and environmental costs. This necessitates the embedding of measurement by matching within a value framework as in our stewardship model.

We are not alone in linking technical finance with stewardship. Entrustedness requires that mangers control and *account for* resources. Block articulates the ethical significance of procedures to spend and to control money. The right to manage others' money "is the operational definition of power in our institutions."[82] This is because "Financial planning and controls are a major device for living out our intentions."[83] As long as they are not reduced to a mere control function, consistent measurement and reporting are of considerable usefulness. The financial functions,

> "help us to become fully informed and communicate about performance. Financial functions also help people, through budgets, to document and keep track of their promises."[84]

These intentions are a "critical means" for people to "fulfil their stewardship responsibility."[85] Although Block takes this stand, he has not worked out a detailed model to integrate technical finance and stewardship ethics. In this respect our work is original.

We cannot proceed without summarizing the technical matching that enables managers and employees to effectively control assets to achieve the firm's goals. This matching is fundamental to our measurable entrustedness and is encountered in business as follows.

Matching Assets and Liabilities

Assets (items owned) and liabilities (items owed) are presented on the balance sheet, according to the accounting equation:

$$\text{Assets} = \text{Liabilities} + \text{Net Worth}$$

to show the financial position. If the net worth is positive the firm is solvent.

Matching Income and Expenditure

Survival necessitates adequate surpluses to cover reducing asset quality and net profit to cover distributions and expansion to enhance the asset value. It is thus necessary to match income (revenue) and expenditure plus depreciation.

Matching Resources and Goals

Matching is particularly important in the planning exercise. Correctly estimating the firm's potential and opportunities, and deploying resources appropriately is essential. This involves four stages in long term planning. The *assessment* stage is a position audit taking stock and answering the question, "Where are we now?" The second stage addresses the firm's *aims* or objectives. The question here is, "Where do we want to be?" Objectives are more complex than just profit maximization. Firms seldom isolate a single goal. Drucker suggests setting quantifiable objectives for: Market standing, Innovation, Productivity, Physical and financial resources, Profitability, Manager performance and development, Worker performance and attitude, Public responsibility.[86] The third stage is to consider *alternative strategies* by which the objectives can be achieved. Essentially this is concerned with answering the question, "How do we proceed from where we are now to where we want to be?" The final stage will be the *adoption* of the best strategy to meet the expectations of the investors.

Matching Cash Inflows and Outflows

Planned income is not the same as cash inflow and planned expenditure is not the same as cash outflow. In both cases there is the possibility of a time lag between the liability arising and payment being made. There will also be other cash outflows to consider such as loan repayments, taxation or additional capital expenditure.

Matching Profit with Investment

Management must make the best use of resources and need to appraise alternative investment opportunities. Investment decisions involve the choice of the best alternative out of several opportunities that will be beneficial. When we limit our consideration of benefit to the financial benefit to the firm, what is involved is an increase in the net worth of the business. By matching the net income of a project with its capital outlay the manager can determine whether or not the project will result in a positive net value.

Matching Actual and Planned Expenditure

There can be wide gaps between targets and achievements and some kind of control system is needed to monitor income and expenditure and compare the actual figures with those planned. Changes resulting from inefficiency can be identified and remedial action taken.

Matching Current Assets and Current Liabilities

Each project requires working capital to operate the fixed assets involved. As the business proceeds however the working capital of the different projects cannot always be distinguished and the position for the whole firm must be monitored.

Block does have some practical suggestions for handling money in ways that reflect a commitment to stewardship. Key points of his ethical stewardship agenda are as follows.

(a) Full disclosure of relevant information.
(b) Economic literacy to enable economic realities to be understood, including the economic consequences of choices made.
(c) Budget accountability for the costs of activities undertaken. Within a stewardship context: "A budget is a promise. It is the commitment to meet a set of objectives, to deliver a set of products or services for a predetermined and fixed level of expenditure."[87]
(d) A monitoring process that identifies the people responsible for economic stewardship.
(e) Auditing for learning and consultation as well as policing and protection. [88]

These key points of Block's agenda reveal his persuasion that stewardship values can be applied to the operational level in a way that improves the organization's sensitivity to values. He has dealt mainly with governance and human relations providing some principles for financial transactions without seeking work these through in detail at an operational level. This supports our idea of embedding financial matching within ethical parameters in our stewardship model making it workable in business operations.

In the following chapters we proceed to apply the concept to borrowing and lending. My own career expertise in accountancy and lending originates in several years' employment in banking followed by lecturing in accounting. Previous business ethics research interested me in the operational potential of the stewardship concept and in the remainder of this book I will be building ethics into this operational world familiar to me. We have demonstrated that stewardship is much more than technical financial matching where costs and benefits are quantified in monetary terms. There is a qualitative moral content, which can have financial implications, but needs to be differentiated from these. Our approach to ethical debt management differentiates technical and moral values and *then* integrates them into our model giving both their proper role so that neither usurps the place of the other; and so that our analysis can do justice to both *and* the interactions between them. Thus, while recognizing the necessity and centrality of measurement by matching in connection with stewardship activity, we emphasize that this criterion must be set in its moral context. By doing this we exclude a "profits by any means" approach. Our intention, therefore, is to examine financial practice at an operational level by our integrated model.

In this chapter I have shown that: (a) technical definitions of stewardship are morally inadequate, (b) my stewardship model can stand up to a variety of criticisms, (c) the stewardship concept when ethically analyzed furnishes a set of values which are capable of application to the operational level of organizational activity, and (d) the stewardship model is truly dynamic, technical financial matching can be embedded within it, and interactions between technical and moral values explored. We introduced debt management scene in *Chapter 1* and in the following chapters show how my concept of stewardship can be worked out in its operational activities.

Notes to Chapter 2

1. T. Lucey, *Management Accounting*, p.1. Tony Hines, *Foundation Accounting*, pp. 1-8.
2. Richard Mattessich, 'Paradigms, research traditions and theory nets of accounting'. In M.J. Mumford and K.V. Peasnell (eds.), *Philosophical Perspectives on Accounting: Essays in Honour of Edward Stamp.*
3. *Ibid.*, pp. 192-199.
4. Kamal H.M. Naser, *Creative Financial Accounting: Its Nature and Use,* p.3.
5. Barry Elliott and Jamie Elliott, *Financial Accounting and Reporting,* p.46.
6. Michael Bromwich, *Financial Reporting, Information and Capital Markets,* p.309.
7. Cheryl R. Lehman, *Accounting's Changing Role in Social Conflict*, p.20.
8. *Ibid.*
9. Richard Mattessich, *op. cit.*, p194.
10. *Ibid.*, p.195.
11. *Ibid.*, pp.195-199.
12. Littleton and Zimmerman, quoted in Cheryl R. Lehman, *op. cit.*, p.19.
13. See Peter F. Druker, *Managing in Turbulent Times,* pp.108-113.
14. John Donaldson, *Business Ethics: A European Casebook,* p.248.
15. *Ibid.*, pp.252-253.
16. *Ibid.*, p.253.
17. Marianne M. Jennings, *Case Studies in Business Ethics*, p.107.
18. Tom Cannon, *Corporate Responsibility*, p.73.
19. John Humble, *Social Responsibility Audit : A Management Tool for Survival*, p.2.
20. *Ibid.*, p.3.
21. Milton Friedman, *Capitalism and Freedom*, University of Chicago Press, 1962, p.133, quoted in John Humble, *op.cit.*, p.1.
22. Steven N. Brenner and Earl A. Molander, 'Is the ethics of business changing?' *Harvard Business Review*, January-February, 1977, reprinted in *Ethics for Executives: Part II*, p.34.
23. Kenneth R. Andrews, 'Can the best corporations be made moral?' *Harvard Business Review*, May-June, 1973, reprinted in *Ethics for Executives: Part II*, p.4.
24. *Ibid.*, pp.4-5.
25. Theodore Levitt, 'The Dangers of Social Responsibility' in Tom L. Beauchamp and

Norman E. Bowie, (eds.) *Ethical Theory and Business*.

26. *Ibid.*, p.138.
27. *Ibid.*, p.139.
28. *Ibid.*
29. Tom L. Beauchamp and Norman E. Bowie (eds.), *Ethical Theory and Business,* Fourth Edition, p.2.
30. *Ibid.*, pp.18-19.
31. *Ibid.*, p.23.
32. *Ibid.*, p.22.
33. Ronald H. Preston, *The Future of Christian Ethics,* p.134.
34. Tom L. Beauchamp and Norman E. Bowie (eds.), *op.cit.*, p.4.
35. T.Lucey, *op.cit.*, p.220; Cecily A Raiborn et.al, *Managerial Accounting*, p.673; and Duncan Williamson, *Cost and Management Accounting*, pp.481-482.
36. Milton Friedman, 'The Social Responsibility of Business' in Tom L. Beauchamp and Norman E. Bowie, (eds.) *op. cit.*
37. *Ibid.*, p.136.
38. *Ibid.*
39. *Ibid.*
40. *Ibid.*
41. *Ibid.*
42. John Donaldson, *op.cit,* p.v.
43. Michael Novak, *The Spirit of Democratic Capitalism,* pp.116-122.
44. John Donaldson, *Key Issues in Business Ethics*, p.88.
45. James W. Kuhn and Donald W. Shriver, *Beyond Success: Corporations and their Critics in the 1990s,* p.267.
46. *Ibid.,* p.272.
47. *Ibid.,* p.277.
48. Ronald H. Preston, *op.cit.,* p.91.
49. President George Bush, 'A New Era of Environmental Stewardship', pp.2-3, *EPA Journal,* Sept.- Oct., 1990.
50. Thomas F. Malone, 'Stewardship: Paradigm for the Nth Millenium', pp.2-3, *National Forum,* vol.75, no.1.
51. Morris Miller, *Debt and the Environment: Converging Crisis,* p.4.
52. Peter Selby, *Grace and Mortgage: The Language of Faith and the Debt of the World,* pp. 74-75.
53. Peter Block, *Stewardship: Choosing Service Over Self-Interest*, p.5.
54. John Passmore, *Man's Responsibility for Nature: Ecological Problems and Western Traditions*, p.31.
55. Cornelius Van Til, *The Defense of the Faith*, p.12-13.
56. Cornelius Van Til, *In Defense of the Faith, Volume III : Christian Theistic Ethics*, p.36-37.
57. *Ibid.*, p.46.
58. *Ibid.*, p.44-45.
59. *Ibid.*,

60. *Ibid.*
61. *Ibid.*
62. See also Douglas Vickers, *Economics and Man: Prelude to a Christian Critique,* pp.2, 130-134; Bob Goudzwaard, *Capitalism and Progress: A Diagnosis of Western Society,* pp.211-214; and Gary North, *An Introduction to Christian Economics,* pp.361-376.
63. Harold L. Johnson, 'Can the Businessman Apply Christianity?' *Harvard Business Review*, September-October, 1957, reprinted in *Ethics for Executives*, p.72-73.
64. *Ibid.*, p.75.
65. Christopher D. Stone, 'Should trees have standing? - Toward legal rights for natural objects.' pp.453-460, 463-464, 480-481, 486-487, *Southern California Law Review*, vol.45, 1972, reprinted in Tom L. Beauchamp and Norman E. Bowie, (eds.), *Ethical Theory and Business*, p.565.
66. Lynn White, Jr. "The Historical Roots of our Ecologic Crisis" pp. 1203-1207, *Science*, Vol.155, March 1967.
67. *Ibid.*
68. Arthur R. Peacocke, *Creation and the World of Science: The Bampton Lectures, 1978*, p.277.
69. *Ibid.*
70. *Ibid.*, p.278.
71. *Ibid.*, p.283.
72. See John Passmore, *op.cit.*, p.9.
73. Francis A. Schaeffer, *Pollution and the Death of Man: The Christian View of Ecology*, p.44.
74. Norman L. Geisler, *Christian Ethics*, p.309.
75. See Norman E. Bowie, 'Business Ethics as a Discipline: The Search for Legitimacy' in R. Edward Freeman (ed.) *Business Ethics: The State of the Art,* pp. 28-29.
76. Technical details can be found in John A. Schofield, *Cost-Benefit Analysis in Urban and Regional Planning.*
77. Michael Common, *Environmental and Resource Economics: An Introduction,* p.282-284.
78. John Passmore, *op. cit.*, pp.49-50.
79. *Ibid.*, p.53.
80. See Paul Heelas, 'Reforming the Self: Enterprise and the Characters of Thatcherism' in Russell Keat and Nicholas Abercrombie (eds.) *Enterprise Culture,* pp.72-90.
81. Cecily A. Raiborn, et. al (eds.) *Managerial Accounting,* p.673.
82. Peter Block, *op.cit,* p.138.
83. *Ibid.*, p.137.
84. *Ibid.*, p.135.
85. *Ibid.*
86. See T. Lucey, *op.cit.,* p.76.
87. Peter Block, *op.cit,* p.139.
88. *Ibid.*, p.138-144.

CHAPTER 3

BORROWERS AND BORROWING

3.1. Stewardship

3.1.1 STEWARDSHIP IN EXISTING FINANCIAL MANAGEMENT

The *Jubilee Centre* survey of individuals with multiple debt problems included in-depth interviews. One focus of the interviews was how money problems had arisen. The findings are summarized in *Table 3.1.*

TABLE 3.1. Factors Contributing to Debt in the Order of Magnitude Suggested by the Respondents

	1st Factor	2nd Factor	3rd Factor
1. A sudden reduction in income	14	6	--
2. Unexpected increase in needs	4	--	8
3. Over-spending etc.	11	23	12
4. An emotional crisis	1	--	--

Source; Jubilee Centre In-depth Interviews [1]

The case material collected at the interviews indicates that although the triggers of default were not always under borrower control, core values could have prevented or alleviated problems. Sample cases are discussed below.

Sudden Income Reduction
Examples of this were unemployment, loss of benefit and marital disturbance. In Mr. T.'s case unemployment disrupted the domestic budget.

> "Mr. T's wife wanted them to continue to live at their previous standard and so they spent the mortgage cheques from the DHSS. ...they continued to accrue debts through having work done to their house on credit, and also through using store credit cards."[2]

In this case there was a failure to adjust expenditure to the income loss, a mismanagement of benefits received to alleviate the unemployment problem and persistence in incurring further debts. Financial *matching* was relevant here, as was a sense of *entrustedness* about the funds from the building society to prevent mortgage arrears. The couple needed clear ideas about necessary *developments* for family welfare. We conclude that our stewardship values could have helped here.

Increased Expenditure

Examples of increases were childbirth and unexpected bills. Mrs. E and her husband had had money problems before the childbirth, which led them to apply for a loan to extend their mobile home. This resulted in unmanageable debt.[3] The particular "cause" proved to be the final straw.[4] Poor financial *matching* already existed and made the family vulnerable. The *development* was sensible in itself but this couple needed a value framework that would recognize the place of development and the importance of financial matching within the context of other relevant values. As we shall see later our stewardship concept answers to this need.

Overspending

The third category embraced *overspending and mounting credit commitments* resulting in crisis. The responses show its importance. Mrs. C is cited as a "fairly typical example". She was borrowing by credit card and in addition had commitments to three finance houses, one loan company and two mortgages.[5] She started with credit cards, was "talked into" double glazing on credit, followed with a fitted kitchen and loans from newspaper advertisements. Repayments became irregular and repossession of the house was threatened for £300 arrears.[6] This woman evidently felt that persuasive salesmanship contributed to their problems. The *developments* undertaken were fairly basic by today's standards but the timing was evidently inappropriate and the easy credit increased rather than alleviated the difficulties. The family needed core values capable of withstanding pressures to run unacceptable financial risks while pursuing legitimate goals. Our stewardship values would have brought balance into this situation.

Emotional Crisis

A family death caused Mrs. L to leave post unopened for a year resulting in unpaid bills. A finance house, after receiving no payments for a year, was threatening repossession.[7] This mismanagement was triggered by emotional crisis and underlines the importance of personal values to withstand life crises.

In getting back to normal life after bereavement, support from family is important but so is a sense of responsibility to others. A sense of *entrustedness* with respect to borrowed funds and *accountability* for them would help back to normality so stewardship values could have helped this lady.

Summary

The causes of default are complex. Particular crises are not the complete explanation. Roberta Hanna concludes her section on the causes of debt problems by saying,

> "Overall, the loss of a job, reduced hours, an unexpected pregnancy, divorce, and an emotional crisis would in most cases not have caused debt by themselves. What seems significant is that, in the majority of cases, the individuals were already well stretched and these key factors just speeded up what might well otherwise have happened."[8]

This conclusion shows that debt management is important in withstanding crises. The cases analyzed above show that stewardship values would have helped and we will show how in the remainder of the chapter.

3.1.2. STEWARDSHIP AND ETHICAL DEBT MANAGEMENT

We have seen above that stewardship values are significant in financial transactions, and borrowers need a framework to handle the complexities of debt management. Single-issue approaches to borrowing require little to go wrong for crises to result. Focusing upon family welfare without matching assets and liabilities or income and expenditure soon produces a crisis. Family welfare goals and financial matching must be integrated in a decision model capable of withstanding external pressures and unexpected crises. We will now show how stewardship can be applied to this. There must be an internalization of these values by the borrower to keep things on an even keel when institutional values are not all that they might be. Financial institutions operate as commercial undertakings in highly competitive environments. Profits and activity targets are a high priority. Sales promotion campaigns are intended to be persuasive. If consumer protection legislation is not to become draconian for lending institutions or paternalistic for consumers, a balance must be struck between institutional and individual responsibility. The long-standing principle of *caveat emptor* emphasizes buyer responsibility. As we saw in *Chapter 1, Section 1.3.4* the balance between the parties is very important. The alleged irresponsibility of the banks in over-aggressive credit promotion has to be balanced by individual responsibility for avoiding default. We saw in *Chapter 1, Section 1.3.2* that borrowers can be at fault, as a result of impulse buying, inadequate financial planning and so on.

Peter Block has used the stewardship concept to emphasize individual responsibility in his book *Stewardship: Choosing Service Over Self-Interest.* Block is a pioneer, but I have extended the concept beyond governance to customer relations and the detailed operations relevant to borrowing. Block's main concern is the application of the concept, *within* the business organization, to relationships between management and employees and different groups of staff. On the subject of individual responsibility he emphasizes that there is a balance of power and that people need to make their own choices.[9] The internalization of stewardship values provides a basis for reform within the workplace. Block sees stewardship as a means to creating self-reliance and prioritizing service over self-interest.[10] It involves the recognition of accountability for our actions.[11] It fosters an independence of action as a result of personal reflection and consideration of the necessary courses of action. There is a maturity involved that does not require a close supervisory control by others. The latter fosters dependency upon others and becomes a situation where our thinking is done for us. In relation to internal organization, Block links stewardship with such ideas as survival being in our own hands, defining our own goals and commitment to the community.[12] His book witnesses to his own conviction that stewardship can be operationalised to reform human relations within the business organization if the values are taken on board by management and staff to govern relationships. This would necessitate personal commitment to the principles and individual application of them to the operational context. It is interesting that Block has carried through his analysis in this way and the reader will discern that there are parallels with our own treatment. The starting point is the *concept of stewardship*. Its moral analysis has squarely faced *the economic dimension*. The importance of *the internalization of the values* is emphasized. The principles are affirmed to be *relevant to the operational context*. We cannot deny that this must be an acceptable approach to business ethics in debt management. It places the onus on the individual with respect to that person's actions and requires the individual to face up to the personal responsibility for what happens as a result of those actions. But it does not leave the individual without a relevant framework of reference for moral decision in order to inform personal action. The application of the concept in this book is more specific than Block's treatment. He has addressed governance at a general level not detailed operations. His concern is that within organizational human relations, workers will be self-reliant. We are concerned with debt management and self-reliant borrowers. Block does not provide any detailed operational decision model linking stewardship and economic matching such as we provide in this work.

Summarized below is the way in which this model will be applied to borrowing in this chapter.

Entrustedness. Entrustedness arises in debt management because the loaned funds are entrusted to the borrower. We explore the significance of this for borrowers.

Measurement by matching. Measurement by matching is required in order to operationalise entrustedness. The latter requires responsible husbanding of assets and liabilities and so on, so it will be appropriate to consider issues related to solvency, and the management of expenditure in relation to income.

Effectiveness. Effectiveness in the control of resources is essential to ensure repayment and this necessitates borrower forward planning including self-assessment.

Development. Development is of broad application in relation to family borrowing and we will need to examine influences on consumer expectations and the defining of objectives as these have a bearing upon financial strategies.

Accountability. The trustee aspect of stewardship involves the responsibility to repay borrowed funds, which involves accountability.

Fairness. Fairness is relevant to the relationship involved in borrowing and lending and has implications for the borrower as well as the lender.

The reader needs to be aware that we are dealing with a model which has interrelationships and interactions so, although we will deal with each value in turn, it is important to realize that *the discussion of one value necessarily requires reference to others where relevant.* The values should not be applied in isolation from one another. This is particularly so with regard to matching by which entrustedness is operationalised. Our integration necessarily involves the discussion of different aspects of matching where relevant to a particular value because we have intentionally embedded the two together. Entrustedness as accountable control of resources is operationalised by matching and this matching has to be integrated with, and not isolated from, effectiveness, development, accountability and fairness.

3.2. Entrustedness in Personal Debt Management

We will now show what difference stewardship values can make in connection with personal debt management. Our analysis in *Chapter 2* showed that the essence of stewardship is *entrustedness,* that this value links the relationship perspective and the resource perspective and that making it operational requires matching. We have shown that entrustedness, as the key stewardship value, is a well understood in a number of different contexts. We do not have a problem here in entrustedness being understood but this is not to say that it is *sufficiently* understood and applied. With respect to stewardship, we saw in *Chapter 2* what Robert M. Adams calls an "overlapping consensus"[13], but this is not to say that the *vigour* of the stewardship concept will be necessarily included. The case studies analyzed in *Section 3.1.1.* suggest that borrowers may have a very shallow view of entrustedness with respect to borrowed funds. This bears out Adam's view that common morality is, "not only imprecise but variable and relative to one's historical and cultural context."[14] It is consequently necessary to specify content and illustrate applications to press home the significance of entrustedness. Common morality is no substitute for ethical frameworks such as ours but it provides a bridge to clarify its content. The existence of core values provides a basis for understandings between those of different ethical stances and, as we have already seen in *Chapter 2*, may involve the adoption of some aspects of what is involved in the ethical concept expounded. The problem arising is that the common values remain as vague ideas until there is a clear specification of the ethical implications from a conceptual basis such as Block gives in his book and we give in this book. Adams is of the view that an ethical account in terms of Divine commands and kindness

> "affords the best possibility of finding a transcendent ground of moral obligation that is consonant with the social nature of obligation."[15]

It is certainly helpful, as we have seen in *Chapter 2*, to consider the breadth of the stewardship concept in its religious setting. Paul Marshall, for example, reminds us of its implications in the following way.

> "We are stewards not only of natural things like land, soil, trees, oceans and minerals. We are stewards of *all* things - time, energy, health, organization, family life, work styles, buildings - everything that exists in human life."[16]

He sees this broad entrustedness as of vital importance in decision taking. It is too easy to limit the latter to the costs and benefits that are readily measurable in financial terms. We illustrated this problem in *Chapter 2.* Positivism has trained us to focus upon financial costs and benefits, usually those that are internal. Marshall sees this as distinctly problematic.

> "... we can (and often do) end up consistently making decisions that are really uneconomic: decisions that consistently have greater real costs than benefits, decisions that make us poorer as people while we maintain the illusion that we are 'growing' economically."[17]

This is particularly true with personal borrowing. What from the point of view of the banker is a commercial transaction, is, from the point of view of the borrower, a domestic transaction and consequently very close to the core of personal life. Mistakes in connection with decisions about borrowing leading to default are thus particularly traumatic.

Entrustedness in this wider sense is too easily overlooked in connection with commercial transactions. It should not be too difficult for borrowers to see that entrustedness arises in the loan situation because the lender entrusts funds to the borrower, but when preoccupied with material possessions and arranging payment for them, consideration of stewardship of time, health and family life may be marginalised. Broader entrustedness prompts the borrower to bring the full range of benefits and disbenefits into the decision taking, whether quantifiable in financial terms or not. Morally this is the way it has to be. Further furniture may be a positive benefit for family life and the financial cost worthwhile. On the other hand additional luxuries may prove a drain on energy and the quality of family life if additional overtime has to be undertaken to make loan repayments. Stewardship entrustedness encourages thinking through the ethical issues more widely than would be the case on a merely prudential basis, broadening horizons beyond material satisfaction to include such things as family welfare. This prevents expenditure from being unnecessarily wasteful and helps to avoid those pitfalls that would lead to personal distress as a result of financial problems.

When we turn to matching in making entrustedness operational we are not reducing stewardship to technical finance, but are seeking to give quantification its proper place in ethical decision taking. What we find in debt management is that decisions about broader entrustedness necessarily involve an interaction with financial matching and the other values. The case in *Section 3.1.1.* illustrates entrustedness in its broad and narrow senses. In the broad sense, the birth of the child increased the scope of entrustedness. The extension of the accommodation was a reasonable response to this. But entrustedness also applied in the narrower sense with respect to the existing borrowed funds, and this should have involved regard for and commitment to creditor interests. In the light of the different relationships it would be necessary to integrate the resource implications of the extension by economic matching. Were this done the following would have emerged. (a) Existing problems in matching expenditure against income. (b) The preclusion, in the light of the childbirth, of resolving the problem by reliance on two incomes. (c) The certain deterioration of income/expenditure ratio and also, as a result of depreciation, asset/liability ratio. We conclude that, what may seem necessary in the light of broad entrustedness, must be modified in the light of the economic matching which would reveal that the couple could not maintain effectiveness in repaying loans. This would be unfair to existing and new lenders and ethically justifies a lesser provision for the family. Such analysis does not displace ethics by letting figures *rule,* but improves ethical decisions by addressing both the relationship and resource sides of entrustedness and the broader and narrower scope of entrustedness.

We conclude that in the domestic context, broad entrustedness has to do with total family welfare. This involves material possessions but also health and relationships. There is an ongoing interaction between these different factors. Material possessions can improve the quality of life and be generally beneficial. Securing such benefits may lead a particular family to resort to the borrowing strategy but debt can result in anxiety, ill health and marital breakdown. It is clear that the moral decision whether or not to proceed cannot be properly addressed in the specific practical situation independently of the measurement of financial impacts. Apart from the latter only ethical generalizations are possible. It is only when measurement by matching is addressed that specific decisions about the overall enhancement of family welfare can be reliable.

3.3 Matching

We will now see in what specific ways matching can be of assistance. We must first see the importance of measurement as derived from entrustedness for personal debt management. When entrustedness is viewed in relationship to resources it involves responsible resource management. The term "husbanding" in the sense of economizing in the use of resources emphasizes this. The right decisions must be taken about matching liabilities and assets and the use of the latter. This matching necessity derives directly from entrustedness. Borrowers are using depositor's funds entrusted to them by the bank. Although the legal position is not defined in terms of managerial stewardship, as with company directors; the moral situation is one of entrustedness. In both cases this entrustedness involves vigilance and diligence in financial management by matching to ensure that the third party interests are not damaged. This does not collapse stewardship into technical finance, which is legalistic, materialistic and reductionist in orientation, but gives measurement its proper place regarding resources. Block shares the principle that is elaborated in detail in this book, that the economic facts must be integrated within the stewardship model. He states:

> "Stewardship encompasses concerns of the spirit, but it also must pass the test of the marketplace. It must be practical and economic."[18]

In our analysis of the concept in *Chapter 2* we went to some length to address the economic dimension and to show how this is related to moral decision taking. A business ethic that fails to do this will inevitably remain at an abstract level because as soon as it enters into the operational activities of commercial life it must address the economic realities of viable activity and just distribution of returns. This is relevant to the lending banker and, as we will see in this chapter, the personal borrower. The cost-benefit analysis case study in *Chapter 2* illustrated that measurement by matching is required in order to *operationalise* stewardship. It is relevant to resource management and asset enhancement and fundamental to the ethical decision about whether or not a particular project should proceed. Matching should not be viewed as peripheral to ethical decisions. Just as economic reductionism fails by marginalising ethics, moral analysis concerning operations involving resources fails if it marginalises costs and benefits expressed in financial terms. In specific cases we cannot solve ethical dilemmas and then look at the figures. The latter are part of the problem and thus contribute to the decision taking to resolve it. It is by involving measurement that we arrive at sound ethical decisions about operational matters. As we saw above, the ethical justification of borrowing in particular cases cannot be abstracted from the financial considerations, because what might be right in one situation might be wrong in another, *dependent upon* the financial considerations. Responsible borrowing necessitates a rational approach that takes advantage of market values to match assets and liabilities, costs and benefits and income and expenditure in order to avoid crisis. T. Donaldson affirms the necessity in connection with business decisions of:

> "merging the ordinary canons of economic decision making, of profit maximization and market share, with the principles of basic moral theory." [19]

We need the same in personal debt management so that economic decision taking is circumscribed by ethical parameters and moral decisions are embedded in economic reality. For this reason matching is of central importance in connection with our stewardship approach to borrowing. The necessary information is considered below.

3.3.1. DEFINING THE FINANCIAL STARTING POINT

Financial starting points are expressed in terms of solvency. Insolvency is a position where the realization value of the borrower's financial assets fails to cover liabilities. Otherwise the borrower is solvent. To obtain financial information relevant to ethical decisions about solvency it is necessary to match assets and liabilities. John McQueen, founder of the Bankruptcy Association, begins his book, *What to do when Someone has Debt Problems*, with a chapter on "Being Realistic". In this chapter he recommends at the outset that the person with debt problems should draw up (a) a *personal* balance sheet listing assets and debts and (b) a *personal* income and expenditure account [20] and gives relevant examples.[21]

3.3.2. MEASURING ASSET ENHANCEMENT

Stewardship involves a conserving and developmental stance. A primary method of indicating material growth or enhancement is the measurement of changes in net asset value. The personal balance sheet provides data for the measurement of net worth. Total asset value less external claims against it gives a starting point from which it is possible to measure whether or not enhancement has taken place over a period. Accurate valuation is fundamental because over-statement of asset value at the beginning or end of the period will distort the measurement of the increase in value. Wear and tear erodes asset values and replacement cost involves no enhancement of total asset value. It is necessary to provide for unexpected losses by increasing net worth. This provides a safety margin. This increase represents the minimum development necessary in order to protect *existing* living standards. Further increases secure long-term development and greater family security but it is difficult to achieve and is an intangible benefit. Levels of income and expenditure are major influences on net worth and are considered below. The closer a borrower is to the solvency margin the greater the potential for difficulties if asset values fall and have not been allowed for. Adequate enhancement should be a primary objective to help cope with uncertainty and we address this below. Solvency leads on to adopting a strategic direction where expenditure is matched with income to maintain a standard of living that does not erode the margin of safety

3.3.3. RELIABLE INFORMATION FOR STRATEGIC DIRECTION

With respect to a private individual, sources of income and types of expenditure differ from those of a business, but the necessity of matching income and expenditure remains. This is not to commercialize family life. It is to recognize that resource management is relevant to domestic life. In the case study in *Section 3.1.1.* there was a serious failure to match expenditure to the income following the income reduction and a consequent incurring of greater debt. The couple needed to assume control. McQueen also introduces matching at the outset in handling debt problems [22], and he would doubtless agree that this debt problem might have been avoided if earlier attention had been given to matching income and expenditure. Current practices do not support our stewardship view of matching at this point as the matching format borrowers encounter in the loan application form cannot be relied upon.

Loan Application Form Approach

The way in which the income and expenditure information is presented is important. The income and expenditure presentation most familiar to private individuals would be suggested by the type of financial information requested on loan application forms. These request details of all income and subtract mortgage or rent, life insurance premiums, rates, rentals and other expenditures.[23] Lenders prepare such forms to assess loan proposals by determining the sum available to make repayments. The problem with such an approach is that it fails to distinguish between *invested* savings and *expenditure*. It is necessary to distinguish between payments that result in net asset enhancement by increasing asset values by purchase or investment, or by reduction of liabilities, and those that do not. The latter we will call consumption expenditure as the benefit has little duration beyond the time of payment. We have stressed the importance of solvency measurement by net worth and the need for accurate financial statements to inform ethical decision so it is important for borrowers to distinguish between saved income, which increases personal solvency and consumption expenditure, which does not. If the latter is allowed to exceed income, personal solvency decreases.

Stewardship Approach

Statements based upon loan application forms, where no distinction is made between mortgage, rent, life insurance premiums and local tax convey the impression that all that is important is the *overall surplus* available to service additional loans. This obscures the more important consideration of the extent to which outgoings affect solvency. A distinction needs to be made between rent and local tax, which are consumption expenditure, and life assurance premiums, which are a form of saving and mortgage repayments, which reduce liabilities. With the mortgage payments a distinction is necessary. The mortgage is related to an asset purchase and this asset increases net worth to the extent that its value exceeds the loan outstanding. The loan repayment consists of two elements (a) repayment of the amount borrowed (a capital item) and (b) payment of the interest on the loan (consumption expenditure). The capital repayment reduces the loan outstanding and this reduction increases the owner's equity and consequently net worth. The outgoing is part saving and property investment and part consumption expenditure for the borrowing facility. Two borrowers may be in the same position from the point of view of surplus available to service further borrowing but from the point of view of financial survival their positions may be very different. This is because a £2000 outgoing repaying a mortgage will be classed in the same way as a £2000 payment for a holiday. It reduces the surplus available to make repayments on a new loan. But in terms of financial security the borrower paying £2000 off the mortgage has increased personal solvency by £2000. Relevant matching for good stewardship requires attention to be directed to how the use of the funds affects financial survival and family security.

It is thus of practical importance regarding the provision of financial information for our decision model that the personal income and expenditure account recognizes the distinction between consumption expenditure and invested savings or loans repaid. This necessitates distinguishing between loan capital repayments and loan interest payments and also distinguishing between wasting assets and appreciating assets. Splitting the calculation of surplus income into two stages clarifies the financial situation. This may be done by the calculation of surplus A, which represents income net of tax less all consumption expenditure, followed by the calculation of surplus B, which represents surplus A less all invested sums and loan repayments. This illustrates how stewardship influences the kind of measurement that should be undertaken and the matching which is relevant in the borrowing context. Family stewardship is more complex than most realize and this section has shown that good stewardship requires certain conceptual ideas to undergird financial practice. We can now specify the following.

(1) Stewardship requires measurement by matching to be seriously addressed *within* the ethical decision model to integrate costs and benefits into long range plans for family welfare.

(2) Matching assets and liabilities determines net worth giving a reasonably accurate measure of personal solvency. If the solvency margin is small, a relatively minor crisis might lead to default.

(3) The concept of net worth can be applied to determine the extent of asset enhancement over a period and this is useful information in identifying changes in the personal safety margin.

(4) The method of matching must be relevant to the decisions to be taken. Consumption expenditure does not increase net worth and must be distinguished from savings and repayment of liabilities that increase the solvency margin and represent genuine enhancement.

(5) Stewardship matching encourages discriminating decision taking. Borrowers are influenced by seller's values and lender's methods of measurement but these do not reflect the borrower's standpoint. Loan application forms are designed to address the matter of repayment not borrower solvency which is the key to family security.

It will be useful at this point to consider family stewardship in a more detail.

3.3.4. FAMILY STEWARDSHIP

Although the banker deals with an individual borrower, the person will usually have family responsibilities and others will be affected by decisions taken. We have, for this reason, emphasized the importance of *family* welfare. This subject is a useful bridge between entrustedness, measurement for purposes of meeting information requirements and forward planning to ensure effective stewardship. In dealing with the scope of stewardship B. H. Hwang draws attention to the fact that stewardship extends beyond property guardianship to its use and the impacts of that use upon others.

> "...the principle of stewardship in relation to property takes account not only of the responsibility of stewards for administration, management and guardianship of property, but also the responsibility of right use, disposition and sharing of property amongst others."[24]

Indeed, it is necessary to go beyond this and to recognize that family welfare itself is something with which parents are entrusted and that the stewardship of property fits within this broader stewardship. In the context of domestic stewardship the complexity of the situation needs to feature in the decision taking. There are several family aspects having implications for family finance that need to be addressed.

Dependants and Family Welfare.

In domestic stewardship we must recognize the importance of family welfare. The individual borrower with whom the bank deals belongs to a family unit and the financial affairs of the unit affect the individual and vice versa. Society looks to the family to provide an intimate and immediate support to the individual members. The *Church of England* report *Something to Celebrate: Valuing Families in Church and Society* emphasizes the importance of the family in society.

> "The family has the prime responsibility for the care and upbringing of children and for the care of its own sick, disabled and frail members. In the main, families shoulder these responsibilities willingly and lovingly, and in so doing they make a major contribution to the well-being of the whole society."[25]

It is clear that the kinds of responsibilities that society regards as primarily those of the family do have a bearing upon debt management. The above quotation indicates the complexity of commitments that healthy adult members of a family may be involved with. As we have already seen in *Section 3.1.1.* of this chapter, emotional crises can have major impacts upon financial management and lead to default. Family circumstances have a bearing upon the potential for crisis. Children and the aged are more vulnerable to the unexpected accident or ailment and increase the range of possible adverse occurrences that may have financial implications and, when dealing with points that have a bearing upon family security, these are factors to be borne in mind. Peter Selby draws attention to the way in which the search for family security can itself result in serious *insecurity*.

> "... the sense of physical security provided by house ownership will certainly prove to be an illusion if the owner falls victim to one of the many other insecurities which have become more and more pronounced even as home ownership has been extended. At a time when employment is uncertain ... mortgage repayments are a long-term commitment which can have very serious consequences ... when stability of income is much less assured."[26]

Selby's point underlines the necessity of forward planning that takes a broad view of stewardship and which is not limited to financial assessments. This will be taken up in *Section 3.4.* below.

Several Income Earners, Mutual Dependence and Support.

Taken in isolation, what has just been said about dependants and family welfare could deter bankers from sympathetic consideration of loan applications from key income earners in families. The increased risk and uncertainty arising from the family unit to which the intending borrower belongs, needs to be balanced by the consideration of the increased sense of responsibility that the borrower can be expected to have and the increased support that the borrower might expect from the family if difficulties arise. The cases considered earlier in this chapter showed that although debt difficulties may strain family relationships they may also produce an increased sense of family solidarity and a determination on the part of the members of the family to help one another through debt crisis. In such cases savings or part-time earnings might be mobilized in order to alleviate debt problems and may prove to be sufficient to restore financial matters to equilibrium. Mike Wolfe recognizes that this aspect of support may prove to be important for the Debt Adviser when deciding upon priority debts in the event of default. One of his criteria for priority is "deprivation of place in the community". Of this he states:

> "It should be used to give priority to those debts whose non-payment

> would cause rifts with family or wider community, the effects of which would be to destroy necessary support networks or to damage mental or physical health."[27]

Wolfe identifies a useful point, worthy of more consideration by creditors. Instead of seeing such networks as competitors for repayment, if they are brought into the equation in a genuine partnership, more serious default might be prevented.

The Problem of Poverty.

Wolfe says of his *Debt Advice Handbook:*

> "This book is written on the assumption that debt is caused by lenders who predict it but lend none the less. It should not be blamed upon the individual actions of borrowers. Individuals are led into borrowing by low wages or the paucity of benefit rates and/or the pressure to consume. They are forced into debt by factors beyond their individual control like rises in interest rates or unemployment."[28]

As a generalization this is clearly an overstatement. Although we have already shown in *Chapter 1* the unacceptability of lax lending criteria, we have also found it necessary to underline borrower responsibility. We can agree that there are macroeconomic forces that are beyond the control of the individual, but individuals are not unaware of these problems and we could legitimately expect them to take such factors into consideration. We would also have to say that individuals do have a responsibility to resist pressures to consume; although we would want to balance that by agreeing with Richard Higginson that the irresponsible issue of credit cards, without adequate credit checks, can result in the poorer sections of society being the ones who end up paying high interest sums because they have difficulty making ends meet.[29] But we would have to accept that low incomes and debt are a combination for default. The problem of family poverty is particularly acute in matters of finance. If the budget is already inadequate it will not be possible to demonstrate ability to repay a loan. Having said this, as will be seen below (*Section 3.6.3.*), loans are given, even where budgets are very tight. The banker is incapable of avoiding criticism here. If the manager fails to lend, the bank will be criticized for lacking a social awareness. If the manager lends and default occurs, the bank will be criticized for lax lending criteria. But there is a deeper problem with respect to poverty. As *Faith in the City* pointed out:

> "Poverty is not only about shortage of money. It is about rights and relationships; about how people are treated and how they regard themselves; about powerlessness, exclusion, and loss of dignity."[30]

The idea of poverty as social exclusion would leave us on the horns of a dilemma with respect to debt management for the reasons following. If the family budget is incapable of demonstrating an ability to repay a loan it is obvious that lending cannot take place on the normal commercial basis. If this is the case, lending can only take place under some special scheme. But if the issue is about "loss of dignity" the borrower will not want to be granted the loan on terms that identify the borrower as not personally creditworthy. It may be that bankers can be persuaded to make more loans to those in need in connection with their social responsibility programmes, but it is evident that in the interests of other stakeholders they would have to treat such loans differently, because normal lending criteria cannot be applied. Without doubt, the problem of poverty needs to be addressed in society, but if there is to be any seriousness about social status, which in debt management involves creditworthiness, this can only be tackled by improving *incomes*, not by increasing *lending*. As Peter Selby comments:

> "The statistics of the misery of those for whom debt is the inescapable context of life are a relentless sign of a society which assumes that provided a person can be lent money, that solves the problem."[31]

Selby makes a telling point here. Reliance upon lending to try and solve poverty will exacerbate not improve the situation.

3.4. Effectiveness

In *Chapter 2* the case was argued for recognizing human beings as responsible planners and decision-takers. We now consider how planning and decision taking should feature within the personal stewardship model so that resources are matched with expectations. *Effectiveness* requires that, in connection with expectations, goals are set and achieved. Actions undertaken should be relevant to stewardship aims. Too often debt management in problem situations becomes a matter of being efficient at keeping creditors at bay by maintaining payments without any clear recovery objective. This is an example of an increasing efficiency that is ineffective in resolving the problem. This is often the pattern of default. Stewardship values direct attention to strategic actions in the pursuit of realistic goals. Gilbert W. Fairholm emphasizes the importance of incorporating values in a practical way into the plans and purposes of managers.

> "Leaders cannot expect success if they appeal to followers in terms of vague slogans about equality, liberty, respect for life, and so forth. As they incorporate these values in their visions and actions and use them to arbitrate internal (and external) conflict, they attract and keep followers.[32]

Personal debt management is similar. It is necessary to see how the values adopted work through and influence the practical decision process. The point was made in *Section 3.2.* that entrustedness has a breadth to it that requires merging of financial and personal factors in the decision making process and this is what will be found in this section.

In considering personal financial planning it is appropriate to look at self-assessment, expectations and alternative strategies. In looking at the personal audit we must be mindful of Rob Gray's caution that, when applied to other situations, *audit* does not have the narrow meaning or objective of its original accounting context. In speaking of environmental audits Gray states:

> "... with the term 'audit' having been appropriated from accountants- as was the case with 'social audit' in the 1970s and 1980s- there is a great need to take care with terminology so that one is entirely clear about the objective of the activity undertaken."[33]

As with environmental audits, the purpose of the personal audit is not restricted to verifying a report about performance. The purpose of such a self-audit is to analyze relevant information with a view to aiding an ethical decision process. Just as Gray maintains that clarity in the environmental audit can save " money, time and a great deal of disruption and heartache"[34] the same is true with respect to a personal stewardship audit as the basis for ethical debt management. Finance occupies a key position but not an exclusive position. It is set within the context of the total life situation. We thus consider the borrower's assets, income, potential and constraints and see how matching facilitates effectiveness.

Assets

What we have said about solvency and physical and financial assets shows that the personal balance sheet should feature as part of the personal position audit to disclose the financial strength or otherwise. Assets should, however, be construed more broadly and this brings into consideration educational background, vocational skills and so on as these have income potential.

Income

Reduction of income features in the surveys as one primary cause of debt problems. The *Jubilee Centre Survey* had respondents who identified as the first factor contributing to their money problems such things as loss of job, reduction of hours and so on.[35] In a chapter on routes into mortgage arrears Janet Ford states,

> "...separation leading to reduced income was an important part of the route into debt in a number of cases."[36]

It is too easy to regard income as guaranteed whereas it is vulnerable to change, particularly in connection with marginal income which may depend upon a spouse's part time earnings, upon qualifying for benefits or upon overtime hours. Quite unexpectedly there may be quite *significant* falls in income arising from relatively *minor* changes in circumstances.

Potential

Financial position, education, skills and present income can all enter into a realistic assessment of one's potential. If considering self-employment, perhaps a franchise, skills and finance will be important, whereas, if looking at the possibilities of promotion, education and skills are likely to be more significant. Potential certainly influences a person's expectations and should enter into the total plan for the future.

Constraints

The individual is faced with constraints and these may assume a greater significance if money problems are subsequently encountered. The intending borrower should take stock of age, health, emotional make-up, number of dependants and so on, as the effects of serious debt can be social, emotional and physical, and the borrower must assess whether the family could cope if problems arose. Roberta Hanna mentions five main types of problem identified by respondents with serious debts. They were (1) mental problems, (2) social withdrawal, (3) physical ailments, (4) marital problems, and (5) family problems. Out of 30 respondents who discussed their situations, Hanna reports:

> "Twenty ... told the researcher that they had been very depressed because of the debts. ... Several lived in great fear of a visit from the bailiff, and several others feared that their house would be repossessed... Overall most of the respondents experienced an increase in tension.
>
> Five of the respondents were suicidal."[37]

Hanna reports how financial constraints and embarrassment produced social withdrawal, worry induced physical ailments, in fourteen of the nineteen married couples a great deal of tension was experienced in their marital relationship and family life could be seriously affected as husband or wife grew increasingly irritable. As one father explained:

> "With these debts as they mount you've got other things that come up as a result of it. You get family stress. Things are on top of you and the slightest little thing makes you very angry."[38]

The frustration of this man underlines the importance of taking stock at the outset to avoid, if possible, such a position. The personal audit is a way of weighing up the situation in advance. It directs thinking beyond capacity to repay to a broader range of considerations. It encourages a long-term view of consequences and raises at the outset the need to consider potential risks and future uncertainties. It gives an impression of what might go wrong and what resources, financial or otherwise, might be available in a crisis. The stewardship concept links all of these together so that financial and personal factors can be balanced in arriving at decisions about particular acquisitions.

3.5. Development

Caring resource management involves constructive use of resources. Objectives should be broader than immediate, unrestrained use of resources. Preservation, conservation and development of resources are relevant as well. Positivism has exerted a continuing pressure to measure. Its concern is with quantification and prediction, and, as we have seen in *Chapter 1*, this moves decision taking in a materialistic direction. It is not just that physical objects and financial expenditures are relatively easy to measure and therefore supplant qualitative factors. There is a deeper problem, as Tom Cannon explains,

> "In its more extreme forms, positivism either asserts that moral issues cannot be addressed or should not be addressed when the opportunity existed to substitute observation and prediction. Freed from 'contamination by values' the study or practice of business and economics can deliver a host of benefits to society." [39]

Such thinking results in an emphasis upon growth that can be easily measured and the pressure for growth is responded to in relation to such indicators as exist. As long as these increase, little or no attention is paid to the fact that immediate or long-term disbenefits have been incurred. Environmental pollution and resource depletion serve as examples of how external costs have been ignored when outputs are aggregated to obtain growth indices figures. Our stewardship development emphasizes that we are not merely aggregating outputs measurable in monetary terms to determine material growth. The latter is important, but only as one factor in the overall picture. Stewardship development requires that the disbenefits arising do not negate the material benefits. The disbenefits may be intangible and not objectively quantifiable but it is of no advantage to disregard them. Qualitative factors must be considered alongside of quantitative factors. As we saw above, the acquisition of consumer durables results in material growth, but if this takes place at the cost of family relationships and personal health we would be wrong to equate this material growth with stewardship development.

This has a bearing upon the objectives defined by borrowers. Assessment must be followed by objectives consistent with the results of the personal audit. The private borrower needs to relate personal goals to personal resources in the broad sense described in the previous section and this cautions against a purely materialistic approach to resource management. Andrew Dobson stresses the need for goals that reflect the whole human being.

> "Human-scale thinking must have spiritual content. If we are to move from partial, fragmented, compartmentalized living towards completeness and holistic living, we have to put back what our dominant, industrial-materialist-scientific world view leaves out." [40]

Our view of stewardship development makes a contribution in this respect, but the borrower must have a firm commitment to it because borrowers live in an environment that is very much conditioned by very different values. Prevailing ideas are very materialistic. The marketplace is very competitive. These and other facts result in powerful influences on consumer objectives in unhelpful ways. We argue in this section that responsible resource management requires borrower commitment to stewardship development so that unhelpful external pressures are resisted and realistic objectives defined.

The question may be raised as to the extent to which we can apply *development* to personal borrowers. We can see in the case of business loans that the banker must follow the entrusted funds into the borrower's business to ensure that they will be effectively used in an appropriate economic development so that repayment can be achieved. In connection with the business development of the borrower, the bank's shareholders and depositors have *first* claim on the returns. As financial adviser the banker must also be satisfied that the development financed by the lending will be of benefit to the borrower's business. In terms of social responsibility the bank must be satisfied that social disbenefits do not outweigh economic returns. The degree of involvement of the bank in the stewardship of the funds in the borrower's business can be significant. The bank may, for example, monitor the progress of large projects. In serious cases the bank manager may attend the Board of Directors. Can any such case be made with respect to personal borrowers?

Although we are applying the stewardship concept at an operational level, our understanding of the concept is strategic. We are not limiting it to legal and technical aspects, but have a *moral* understanding of it. As we said above, we understand development in relation to the total complex socio-ecological context, of which we are all part, and to which we should all contribute. We are not restricting our interpretation of development to the narrow sense of *economic* growth but include a broad sense beyond material development. In both respects we want borrowers to have some recognition of what is involved. We want them to face up to the implications of development in relation to the entrusting of funds to banks and onwards to themselves. They are part of a complex context of stewardship that requires them to address the demands of narrow (economic) and broad (social) development. Some cases of personal borrowing do not differ in principle from what has been said in the previous paragraph. The self-employed borrow to spend in their businesses. Personal borrowers are, however, often employees, not entrepreneurs. They still have to be engaged in economic development to secure income and demonstrate an evident source of repayment. Thus a middle manager receives income from his stewardship of departmental resources. Normally personal loans are not directly related to the development activity that provides the income, but there are even exceptions to this. If the middle manager is an essential car user and borrows to buy a car, the personal loan contributes directly to the economic activity that provides the income from which repayment is made.

Business purpose is not an essential criterion for borrowing, but participation in the total process of economic development is. It provides the income to make the repayments. The broad aspects of development are also relevant. The social aspect of development applies to stakeholder interests and benefit in the case of the company, but in the case of the personal borrower it relates to family stewardship involving *total* family welfare. This is relevant to personal borrowing. We would find it reprehensible for a family to run up debts, resulting in a lower standard of living for the children, just to provide the parents with an expensive holiday. But we would take a different view of a loan for a recuperative holiday to enable a return to remunerative work within the process of economic development. We might say that, *legally*, personal borrowers have no responsibility to develop funds on behalf of the providers of them but what we would want to say, *morally*, is that borrowers ought to take into consideration the way in which development applies to lenders and the relevance of that to themselves; and also the total context of development with its implications for them in both its narrow and broad dimensions. Development, as involving economic growth and social responsibility, we regard as relevant to personal borrowers but the application is limited by their circumstances. The personal borrower usually has no resource deployment responsibility for any business development relevant to the lending, and relevant social responsibility does not go beyond the domestic context. The application of the concept of stewardship must recognize these limits and the values of entrustedness and development, appropriately specified, can accommodate this.

3.5.1. RESISTING EXTERNAL PRESSURES

The development goal of total family welfare is the one at which to aim but the influence of external factors can deflect from this. Government monetary policy can adversely impact on debtors by undermining the soundness of the currency or firms may become pre-occupied with technical values relevant to short-term profitability. The case of Mrs. C, cited in *Section 3.1.1.* illustrated the effects of external influences not counterbalanced by a commitment to realistic development goals.

Awareness of Advertising Effects

Advertising might persuade consumers that material benefits are necessary and that the availability of credit ensures minimum difficulty in securing them immediately. Concern is being registered because there are firms who obscure the real cost of borrowing. Mark Boleat comments:

> "The main abuse is intermediaries advertising mortgages at a rate of interest which is not actually charged on the account. Rather, the rate is that on which repayments are based with deferred interest being accumulated and added to the loan. ...
> "... mortgages are not available at these rates and the danger is that people will be misled into taking out a mortgage, the terms of which they do not fully understand."[41]

Such advertising works against stewardship development by encouraging false expectations and borrowers need a value stance to prevent them from being taken in by such advertising.

Influence of Government Policy Initiatives

Janet Ford refers to the disappointments experienced by those who having entered the home ownership market found themselves in default. She states how:

> "in almost every case borrowers entered with optimistic expectations, with a strong sense of achievement and social success, and perceived their likely advancement."[42]

This is not surprising.

> "The expectations about their likely fortunes that buyers have when they first enter owner-occupation are part of the background against which they actually experience owning. In Britain these expectations are now coherently articulated, providing a clear ideology of ownership that has been more and more visibly shaped and disseminated both by lending institutions and by political parties..."[43]

Our stewardship development would not criticize this sense of achievement. Home ownership on a sound financial basis promotes family security and total welfare. The problem is not in ownership but in integrating the development objective and economic matching. The problem with institutional and political promotion of home ownership has been the failure to take a total view of stewardship. Our decision model would provide a guide to the appropriateness and timing of home ownership.

3.5.2. DEFINING REALISTIC OBJECTIVES

A proper basis for goals relates work to consumption and investment. Consumption goods are consumed immediately. Investment refers to asset acquisition. Borrowers' work finances both kinds of expenditure. Mortgages finance home ownership, requiring future work to be committed to repayment. The idea of present possession from future work has been transferred to consumption by credit availability. "Take the waiting out of wanting" is the familiar slogan. Borrowers' future work is thus committed over a range of expenditures. The case in *Section 3.1.1.* showed that overcommitment was associated with multiple debts. Timescales for repayment and the potential risks concerning ability to work during that period should be considered. Loan appraisal focuses upon what can be afforded at the time of the approval but more comprehensive and long-range data analysis is required. This limits immediate satisfaction but does so to maximize family security. The balance between these is an ethical decision which our stewardship model assists, promoting total family welfare, while leaving the borrower free to make informed decisions on the basis of an integration of economic and ethical values.

This is important because mass production produces a situation where the availability of reasonably priced goods raises expectations about possession beyond economic feasibility. Also, seeing others enjoying a particular standard of living raises expectations without reference to the real costs involved. Sometimes others are purchasing interest free whereas in the case of another borrower *more* future work is necessary to balance present satisfaction because of interest charges. The problem has been exacerbated by post-war periods of inflation resulting in repayments being below the loan plus interest in real terms. This raises other's expectations, as when children, having lived in a home furnished by HP, set off on the same course unaware of changed economic circumstances, such as an adjustment in financial markets to give real returns to savers. They experience default whereas their parents did not. We conclude that development must be related to broad entrustedness, matching and effectiveness.

It is also important to analyze objectives carefully. Durables and non-durables can be distinguished. Some durables have income producing potential. We can distinguish objectives as follows: (a) Personal satisfaction involving immediate consumption. (b) Durable product yielding longer lasting satisfaction over a period. (c) Durables yielding longer lasting satisfaction with investment potential. (d) Assets capable of generating income. (e) Assets capable of generating income and capital gains. (f) Asset yielding personal satisfaction, income-generating with investment potential. Appropriate objectives for a particular individual are a personal matter depending on circumstances and decisions have a significant impact on personal wealth, financial security and family welfare. *Table 3.2* presents example objectives. We cannot equate family welfare with financial security although the latter has a bearing upon the former. However, regarding material welfare, an increasing sense of stewardship would tend to prioritize objectives from the bottom of the table upwards, although for low-income families the objectives will focus on immediate consumption more basic than hotel meals. In each case relevant factors must be addressed using our stewardship model.

TABLE 3.2. Comparative Table of Possible Objectives.

Objective	Personal Satisfaction	Durability	Income producing	Appreciating value
To secure a hotel meal	X			
To secure a dining table	X	X		
To secure a computer	X	X	X	
To secure a large house (part of which could be rented)	X	X	X	X

3.6. Accountability

3.6.1. ACCOUNTABLE CONTROL

Entrustedness as "accountable control of resources" introduces accountability's operational implications. Brian Griffiths emphasizes the human dimension of stewardship. The task of wealth creation,

> "is fundamental to our trusteeship of the world" and when we enter into the task in this way we "discover how work is basic to human fulfilment."[44]

This implies an approach to accountability which is not limited to bank risk and profit. There is a human dimension to commercial activity, which is more important than financial and material considerations. The individual life and work which forms the basis for repayment has significance independent of the loan transaction but this seldom comes into consideration. As long as risk taking is at an acceptable level, with bad debts in line with projections, and profits as planned, individual casualties cease to be part of the equation except where part of the debt might be recoverable. For the borrower the situation is far different. Those with no sense of obligation to fulfil promises have no difficulty walking away from unpaid debts. Commitment to entrustedness and acceptance of the responsibility to repay results in a sense of shame at being unable to do so and it will be important to such a person to fulfil obligations entered into.

Accountability as an aspect of trusteeship should influence action. The obligation to repay involves the obligation to take such courses of action as will enable repayment to be made. Uncontrolled expenditure results in default and culpable default would be a breach of trust. The responsible borrower accepts the obligation to control finances to ensure repayment by monitoring income and expenditure to keep them on target, recording actual expenditures, comparing planned and actual expenditures and identifying and investigating variances. This provides some opportunity for remedial action before things get serious so we will consider domestic budgeting.

3.6.2. DOMESTIC BUDGETING

Accountability requires timely comparison of planned and actual income and expenditures so that timely remedial action can be taken. A statement in columnar form recording planned and actual expenditures side by side for each expense category would suffice. Further columns could record the differences and reasons for them. Monthly recording identifies developing problems and ensures that remedial action prevents cash crises by retaining flexibility to respond to changing circumstances. Preparing domestic budgets is not straightforward for financially unsophisticated borrowers as the empirical evidence shows.

3.6.3. MASTERING BUDGETING PROBLEMS

There is empirical evidence suggesting that domestic budgeting is problematic. The *NCC survey of consumer credit, 1979* identified money worries by household income. Of those with low incomes 25% worried very often. The percentage worrying abated with increasing incomes.[45] The survey results suggest that the outcomes of budget management are related to the available income and that the lower the amount of money coming in the first place, the higher will be the percentage who experience money worries. However, analysis of income support claimants has shown that the low-income group is by no means homogeneous. The *PSI survey of supplementary benefit claimants* showed that 50% of claimant couples with children ran out of money and about 40% of claimant lone parents did the same.[46] So, although income is an important influence on personal budgetary management problems, circumstances must also be considered. Berthoud recognizes that investigators have too easily assumed that the poor are bad budget managers but also that other investigators have been too eager to assert that this is not the case.[47] He goes on to say,

> "Some people find it difficult to manage even on a high income; some keep a grasp on their budget even when their income is low. It is not possible to determine which of the two factors is more important: success or failure depends partly on the circumstances in which people are placed, and partly on how they respond to those circumstances."[48]

The essential factors thus include: income, expenditure, circumstances and decision.

Janet Ford identifies three possible outcomes of the household's financial position "based on the assumption that they will all have to attempt to provide for shelter..., fuel, food, and clothing..." which she refers to as "the basics". She categorizes outcomes, at the point of entering a mortgage, as follows.

> "1. Income insufficient to pay the basic commitments on time.
> 2. Income matched closely to the basic expenditure.
> 3. Income some way in excess of the basic expenditure. Here two further outcomes can be envisaged. One where householders save some, or all, of the disposable income and a second where there is no saving but rather expenditure on goods outside the basic range."[49]

Of the cohort studied by Ford it is disturbing that two or three borrowers had been in Category 1. In one case over-optimism regarding overtime earnings remained unchallenged by the building society and the couple started out with an unrealistic budget where income did not cover basic expenditures.[50] Category 2 accounted for most initial positions and a small number were able to choose between 3(a) saving and 3(b) spending beyond the basics.

Ford gives an example of one single lady who took a £10,500 mortgage with only a £20 surplus.[51] This budget tightness left little scope for correcting adverse variances and there was no scope for reducing expenditure. Anthony Sharp, Deputy Director, *Consumer Credit Trade Association* commented upon this kind of problem in a paper entitled, "Granting Credit - Controlling the Risk". David Andrews reporting on the paper in *Credit Management* summarizes as follows.

> "One of the dangers of borrowing on credit ... was that too many people, in their bid to keep up with the Joneses, overstretched themselves and allowed no safety net to fall back on. "[52]

We conclude that the empirical data suggest that there is a long way to go on the part of institutions and individuals with respect to budgeting. Until this is addressed, as we suggest, problems with accountability will remain.

3.7. Fairness

In the next chapter we consider the relationship between banker and borrower which will introduce the need for fairness. We can conclude at this point that the borrower will be better placed to be fair to lenders if our stewardship model has been used to clarify issues and secure sound ethical judgements through an integrated approach.

3.8. Strategic Decisions for Contemporary Issues

The previous sections have shown that my concept of stewardship will suggest a particular approach to borrowing using entrustedness, measurement by matching, effectiveness, development and accountability. We will now show the kind of strategic decisions it answers to.

3.8.1. STRATEGY SELECTION

John Donaldson heads his list of procedures for improvement with "Matching values to strategies." He details the case of a company needing to expand through increased labour productivity. This seemed unpromising but management and unions started from principles, separating out the factual, prudential and ethical and secured increased output, profit, pay and productivity [53] showing that improvement can be secured through reconciling values with strategies. In this sub-section we will consider strategy using our stewardship values.

Attitudes to Alternatives

Having prioritized objectives alternative strategies to achieve them must be considered. Daniel R. Gilbert, Jr. explains how a concern for purpose can emphasize the alternatives, or possibilities, that we can choose to pursue; it can call attention to the worth of our pursuits acting as a standard, enabling evaluation.[54] In family purchasing strategy there are two main questions: whether to borrow and, if so, from whom? The two basic strategies are: to defer purchase and accumulate the purchase price from future income, or to purchase immediately, securing immediate satisfaction through loan finance, repayable from future surplus income. Repayment involves the capital and interest.

Attitudes towards these two strategies are found in the NCC (1980) survey. Summarized results are in *Table 3.3.*

The majority of respondents (69%) approved of credit, although a significant proportion (29%) was positive to the saving strategy. The results suggest reserve about regular debt dependence; 72% did not want to rely regularly on credit and only a quarter (26%) accepted credit as an everyday matter. Most respondents would keep clear of it if they could. However, 43% were prepared to use credit as an alternative to the saving strategy for essential expensive items. Attitudes to credit (excluding mortgages) are, therefore, more cautious than we might expect.

TABLE 3.3 Summary statements of attitudes to credit [49]

Statement	*% agreement*
A sensible way of buying	6
A convenient way of buying	20
Occasionally necessary	43
Never a good thing	29
Don't know/None of these	2

The total picture suggests that our stewardship approach to debt, subordinating immediate material satisfaction to broad *entrustedness* to avoid risking total family welfare, would interest consumers and help decision taking about whether to borrow.

Reservations about Borrowing

Berthoud summarizes reservations about reliance upon credit. Credit is sometimes regarded as intrinsically bad. It carries with it the risk of a debt that has become a burden. It may also,

> "weaken people's perception of the link between income and consumption and so encourage irresponsible buying."[56]

Other concerns centre on the ease of obtaining credit and the excessive energy of its promotion. These concerns are often associated with the idea of folk on low incomes being prompted to purchase luxury items beyond their means. Further worries about credit have to do with loan terms and conditions. The borrower might not understand the agreements or appreciate the risk involved in mortgaging a home. The cost of the borrowing might be excessive.[57] The list shows a distinction between objections to credit as a matter of principle and those concerning the abuse of credit, irresponsible rushing into or irresponsible pushing into borrowing. The first questions would be, Is credit morally wrong under all circumstances, or just in some circumstances, or in none at all? The second set of questions would have to do with the way in which credit is promoted and arranged and the fact that legislation already exists in this area shows public concern.

Recognition that all Credit involves Debt

Individual strategy formulation requires an acceptance of the relationship between "credit" and "debt". Berthoud summarizes the popular conception that credit becomes debt when repayment difficulties have arisen.

> "'Debt' has been defined as the credit transactions which people have come to regret, because they cannot easily repay the money."[58]

On this understanding someone with an overdraft would be able to say that they were not "in debt". This view removes the stigma of being a "debtor", but it is not the *commercial* reality. Bankers would soon take fright if overdrawn customers were to believe they did not have *debts*. Where money is owed there is a debt and debtor. The concept of *credit* relates to the debtor's reputation concerning intention and ability to repay. The credit *limit* reflects the lender's perception of the borrower's honesty and financial standing. In the transaction, however, the borrower is debtor not creditor. The lender "gives credit" but by taking that credit the borrower goes into debt. This does not contradict the *Crowther Committee*'s conclusion that "on balance consumer credit is beneficial"[59], it merely emphasizes that consumer perception of credit should be comprehensive. Janet Ford refers to an attempt to perceive debt in a more positive way in Seligman's presentation of credit as enforced saving or a way of "storing up future utilities."[60] This is more helpful for mortgages than consumer debt but in both cases third party claims against present assets and future income remain.

The idea of indebtedness helps us answer the question, Is credit morally wrong in all circumstances, in some or never? Indebtedness focuses attention on the spectrum of personal circumstances in relationship to third party claims. At one extreme are debtors whose assets dwarf their debts and whose income dwarfs the repayments. At the other extreme are debtors without assets and with repayments in excess of income. With respect to the former *some* would still say credit was bad, but with respect to the latter *few* would deny that it is a bad situation to get into. Credit use can be convenient and our stewardship model can clarify the ethical decision in the light of all the circumstances.

Those forced into borrowing at high interest rates to finance basic necessities are a specific problem. The consumption is debt dependent and cannot be resolved merely by application of our stewardship model. The borrower has lost control and needs external assistance. This may be debt forgiveness, which is discussed later, or charitable support. The *effectiveness* we outlined can maintain an improved position, but only external assistance could remedy such default where personal income is already inadequate to cover necessities.

The Problem of Uncertainty

Defaulting debtors may not have neglected financial responsibility. Credit is given in an environment of uncertainty and unexpected circumstances upset plans. The stronger the impressions of uncertainty the more one would be disposed to the savings strategy, whereas the stronger one's impulses to present satisfaction, or the more pressing one's necessities, the more one would be disposed to resort to borrowing. However, the borrower who is in difficulties through obtaining credit to make ends meet will secure more sympathy than the defaulter who borrowed for non-essentials. In all cases addressing *development* in the light of *accountability* for borrowed funds will lead to sounder decision taking. The second set of questions concerning credit promotion will be dealt with in the following chapter, which addresses lending practice.

3.8.2. STEWARDSHIP OF DEBT

Structured or Flexible Borrowing?

The relative significance of different providers of personal finance is shown by the results of a *Jubilee Centre* survey in 1986/87 based on 1,043 cases.[61] The sample is of debtors who sought money advice through an agency. Although Social Classes I to V were included, 43.1% of the respondents were unemployed and 30.65% were employed but their social class was not known. The findings are summarized in *Table 3.4.*

TABLE 3.4 Type of Debt of Total Sample (excluding Liverpool)[62]

Type of Debt	*Average Debt(£)*	*%*
Finance Companies	1954	42.8
Banks	1030	22.6
Credit Cards	349	7.6
Other Debts	375	8.2
Stores (cards etc.)	306	6.7
Rent/ Mortgage arrears	290	6.3
Provident, Mail Order, etc.	263	5.8
Total	4567	100.0

Looking at consumer credit default excluding mortgages, the ranking in *Table 3.4* shows the importance of lending by finance companies and banks. Credit cards and store credit are less significant.

The *Department of Trade and Industry* comparison of sources of consumer credit in 1976 and 1987 [63] showed that in 1987 lending was greater by the following approximate multiples:

Finance houses	4 times
Banks	1.7 times
Credit cards	6 times
Retailers	Static

This information shows the trend to greater revolving credit use. What may have kept credit cards at such a low percentage in the *Jubilee* survey is the high proportion of unemployed in the sample. Empirical material for the same period indicates the limited use of credit cards by low-income groups. The *OFT survey of credit practices, 1987* [64] related sources of credit used to the income of users and found that those on lower incomes were less likely to use credit cards and loans and more likely to use mail order. Credit cards, loans and store credit are used to a greater extent by wealthier folk. However, the less well off were represented in each credit source and it would appear that there is some access to the different sources of credit irrespective of income. The Office of Fair Trading *Moneyfax* guide to credit and debt draws attention to unrequested credit limit increases as a route into overspending.[65] The high percentage of debts to finance companies in the *Jubilee* survey prevents the simplistic conclusion that debtors likely to default will be safeguarded by using regular installment finance as opposed to some kind of revolving credit. Although credit cards do present additional dangers it does not appear that structured as opposed to flexible borrowing will be decisive or necessarily significant in preventing default. The situation is more complex. Although selection of the type of financing is important it appears that this choice will not provide safeguards independent of personal stewardship. Borrowers must accept responsibility for their own debt management and not rely on lenders to keep them out of trouble. The following specific matters show why our stewardship model is so important.

Effective Debt Management

Complexity of Causation. Janet Ford argues that it is not enough to latch onto some reason for default in isolation from the broader picture. Descriptions such as ‘unemployed’, ‘low income’, ‘single parent’, etc., are best regarded as "likely predispositions". These groups have least margin between income and expenditure, and this

> "renders them particularly vulnerable to financial problems, initiated or made more extreme by some change of circumstance(s) such as unemployment, marital dispute, illness, or pregnancy. Other groups may be vulnerable because of over-commitment on the back of easy credit, reliance on credit because of poverty or miscalculation of costs (especially housing maintenance and heating costs), all of which may create problems for the budget."[66]

In response to these comments note that we have already argued that stewardship requires assembling reliable information for strategic direction including the *matching* of income and expenditure (*Section 3.3.3.* above) and resisting external pressures by appropriate *development* goals (*Section 3.5.1.* above). The importance of the domestic budget was addressed under *Accountability* (*Section 3.6.* above). The complexity of causation emphasizes the importance of a comprehensive model for personal debt management.

Over-extended on Mortgages. The 1990s revealed that over-commitment is not a problem exclusive to low income families. The *Skipton Building Society's* home repossessions in 1990 were concentrated in occupations normally associated with high incomes with accountants, clerical, architects, and doctors heading the list. More than half of the repossessions were in the South of England.[67] Terry Adams the Skipton chief executive explained that this arose because,

> "professional people ... thought that prices could only go up and - when the abolition of multiple interest relief was announced and interest rates were much lower - they extended themselves further than they would normally have done." [68]

This is an interesting case because the crisis is related to strategic decisions that would be approved by one aspect of our stewardship model. We argued for the prioritization of objectives of long-term asset acquisition yielding, not only personal satisfaction, but asset enhancement through appreciating values. The default of over-extended owner-occupiers demonstrates the importance of using a comprehensive decision model so that all factors are taken into consideration. We argued the centrality of family welfare and security in our stewardship model (*Section 3.2.* above) and the necessity of matching of assets and liabilities and liabilities and ability to repay, in individual cases, so that appropriate margins are maintained to cope with uncertain future events.(see *Sections 3.3.1-3* above).

Macro economic Influences. In an article entitled *The Rise and Fall of House*

Prices: Causes, Consequences and Prospects M.C. Fleming and J.G. Nellis analyze the causes of the sudden acceleration in the rate of house price inflation in 1988.(69) They identify the following: (a) Deregulation resulting in credit expansion as lending restrictions eased, new players intensified competition and more flexible lending criteria were employed. (b) A sharp decline in interest rates engineered in relation to exchange rate policy. (c) The pre-announcement of the abolition of multiple mortgage tax relief creating a rush for mortgage funds to beat the deadline. (d) The reduction in personal income tax rates resulting in an increase in disposable income. (e) The influence of the "wealth effect", which the authors state,

> "is associated with the growth in owner-occupation and with the extent to which the value of the owner-occupied stock of houses had more than kept pace with inflation throughout the past two decades or more. The figures show that personal sector gross wealth (including dwellings) more than doubled in real terms over the period 1969-1988; excluding the housing element, however, brings the growth down to an increase of around two thirds. Against this background, people's expectations of prosperity and future affordability of higher mortgages were enhanced still further by the perception of an optimistic economic outlook prevailing in mid-1988."[70]

The boom was unsustainable. The 1988 budget stimulus was a one-off and the problem was compounded by rising interest rates increasing borrowing costs.[71] Prices and turnover in the property market fell and for those in trouble selling up became difficult. In previous sections we have emphasized the importance of core values to resist external pressures such as government policy initiatives which sweep consumers into unrealistic expectations (*Section 3.5.1.* above). The importance of realism in matching work and consumption (*Section 3.5.2.* above) also applies to investment. Family security demands that the rigorous decision process outlined above (*Sections 3.3., 3.4. and 3.5.*) are not bypassed. Borrowers would have been safer applying our model.

Relaxation of Lending Criteria. Reflecting upon problems of default Mark Boleat admitted that the notice of withdrawal of multiple mortgage interest relief precipitated a rush to borrow which led to "some cases" in which "lending criteria were not tight enough." Boleat maintains, however, that the majority of the arrears happened because of changed market conditions.[72] Lending policies were affected. Maximum mortgages of 100% of valuation fell to 95% and multiples of 3.5 times single incomes fell to 3 times; but for the casualties these modifications came too late. The point was underlined that borrowers should not depend on lenders regarding decision taking. Lenders make mistakes. The assessment of changing market conditions is no easy matter. Lenders survive default because their risk is spread. Individual borrowers have no such protection. Their whole future is tied into the debt. Our model would have provided protection.

Borrowing out of Debt. Having made bad borrowing decisions some defaulting borrowers try to avoid the day of reckoning by greater dependence upon debt. Stewardship as measurable entrustedness links the repayment obligation and financial matching to ensure loan servicing and capital repayments. Thus the focus should be upon budgetary control but often people try and borrow out of debt instead. A survey in 1990 by *Signal International*, entitled *Life after Debt*, found that 25-34 year olds spent an average of £919 a month and 35-44 year olds £831. Approximately 28 per cent of the expenditure came from borrowed funds. Together, these two groups account for one third of the population but one half of the outstanding debt.[73] Unfortunately the institutional provision of appropriate loan packages encourages consumers to seek to solve problems by further debt.[74] The *National Remortgage Corporation* advertising describes a remortgage as follows.

> "It is simple and straightforward - it means we arrange to swap your existing mortgage to a new lender (either a bank or building society) normally at a lower rate of interest and also - if required raise up to 85 percent of your current house value. After paying your existing lender you can then use the balance for any purpose..."[75]

It is not surprising that those experiencing repayment problems consider remortgaging. Negotiating a new loan through a broker saves having to face existing creditors.[76] Deferred rate schemes remove the immediate problem but at the cost of later crisis. In "A mortgage fix to soothe that repayment headache" the impact of a *Hill Samuel Mortgage Services* deferred rate mortgage is explained.

> "For someone with a £50,000 mortgage opting for the full 5 p.c. deferral, the monthly repayments would be £234 in the first year." By the sixth year, when the fixed rate period would be over, the monthly payments would have "more than doubled to £552." The borrower would still "be paying 2.5 p.c. less than the variable rate." The debt would have increased to £61,847.[77]

We have presented a model that addresses long term fundamentals not such a remortgage "quick fix". Our stewardship debt management model offers a far more promising future, even if, in the short term, there is no quick or painless solution.

Credit Promotion Emphases. Glossy brochures emphasize present enjoyment not financial survival. "Get a Barclaycard and get away from it all with a

FREE Golden Key Weekender Break."[78] The pictures of conservatory or home extension focus on life-style.[79] Text reinforces visual impressions about freedom to spend as one would like [80] to secure improved living standards without fuss or worry [81] Freedom from concern sounds ideal but things may not work out that way. In London inquiries to *Citizens Advice Bureaux* about consumer and debt issues rose over 40% in an 8-year period. In some bureaux up to half of these related to debt.[82] Blame is not exclusive to advertisers. Many borrowers experience trouble-free borrowing and some brochures mention financial risks. The case studies of default witness to a lack of attention to financial survival as one cause of debt problems. Entering into debt without a debt management model such as we have presented is *likely* to lead to trouble. If borrowers go through the steps itemized in *Sections 3.3., 3.4* and *3.5* they would at least be better prepared for the unexpected.

We have now shown that the stewardship concept can be applied to personal borrowing and that its values are relevant and useful in determining practical recommendations for ethical debt management. We will turn in the next chapter to apply the concept to the operations of lending institutions.

Notes to Chapter 3

1. Roberta Hanna, 'Families in Debt: The People and Their Money Problems'. In Andrew Hartropp, (ed.), *Families in Debt: The Nature, Causes and Effects of Debt Problems and Policy Proposals for Their Alleviation*, p.44: copyright (c) 1988 Jubilee Centre Publications Ltd.: reproduced here by kind permission of Jubilee Centre Publications and Roberta Hanna.
2. *Ibid.*, p.46.
3. *Ibid.*, p.48.
4. *Ibid.*
5. *Ibid.*, p.49.
6. *Ibid.*
7. *Ibid.*, pp.49-50.
8. *Ibid.*, p.50.
9. Peter Block, *Stewardship: Choosing Service Over Self-Interest,* p.xx-xxi.
10. *Ibid.,* p.5.
11. *Ibid.,* p.6.
12. *Ibid.,* pp.33-40.
13. Robert M. Adams, 'Religious Ethics in a Pluralistic Society'. In Gene Outka and John P. Reeder, Jr. (eds.), *Prospects for a Common Morality*, pp.93-94.
14. *Ibid.*, p.96. C. Whyley et. al. *Money Matters* (1997) p.95 reported finding five overall approaches to money management: "moral planners, pragmatic planners, flexible planners, muddlers and pay-as-you-go."
15. Robert M. Adams, *Ibid.,* p.96.

16. Paul Marshall, 'Towards a Christian View of State and Economy'. In Jonathan Chaplin (ed.) *When Christians Disagree: Politics and the Parties,* p.47.
17. *Ibid.*, p.48.
18. Peter Block, *op. cit.*, p.xxii.
19. Thomas Donaldson, *The Ethics of International Business*, p.101.
20. John McQueen, *What to do when Someone has Debt Problems: A Practical Survival Guide*, p.15.
21. *Ibid.*, pp.16-17.
22. *Ibid.*, p.15.
23. TSB Preferential Loan Application form CBSD26/IFA.
24. B-H Hwang, *The Theology of Property: a Consideration of Christian Responsibility of Ownership and Use in the Bible and Tradition,* Cardiff University: Ph.D., 1995, p.99; reproduced here by kind permission of B-H Hwang and Cardiff University.
25. Church of England, *Something to Celebrate: Valuing Families in Church and Society*, p.159.
26. Peter Selby, *Grace and Mortgage: The Language of Faith and the Debt of the World,* p.38.
27. Mike Wolfe, *Debt Advice Handbook,* p.119. See also pp. 122, 154-155.
28. *Ibid.,* p.3.
29. Richard Higginson, *Called to Account,* p.111.
30. *Faith in the City*, Church House Publishing, 1985, quoted in Richard Higginson *ibid.*, p.157.
31. Peter Selby, *op. cit.,* p.50.
32. Gilbert W. Fairholm, *Values Leadership: Toward a New Philosophy of Leadership,* p.66
33. Rob Gray et al., *Accounting for the Environment*, p.101.
34. *Ibid.*
35. Roberta Hanna, *op. cit.*, p.44.
36. Janet Ford, *The Indebted Society: Credit and Default in the 1980s*, p.106.
37. Roberta Hanna, *op.cit.*, p.55.
38. Quoted in Roberta Hanna, *ibid.*, p.57.
39. Tom Cannon, *Corporate Responsibility*, p.99.
40. Andrew Dobson, *The Green Reader,* p.104.
41. Mark Boleat quoted p.2 *Savings and Loan News*, July, 1989.
42. Janet Ford, *op.cit.*, p.87.
43. *Ibid.*, p.86.
44. Brian Griffiths, 'The Conservative Quadrilateral'. In Jonathan Chaplin (ed.) *When Christians Disagree: Politics and the Parties*, p.244.
45. Richard Berthoud, *Credit, Debt and Poverty*, p.5.
46. *Ibid.*, p.6.
47. *Ibid.*, p.5.
48. *Ibid.*, p.6.
49. Janet Ford, *op. cit.*, pp.80-81.

50. *Ibid.*, p.81.
51. *Ibid.*, p.82.
52. David Andrews, 'Dealing with debt', p.41 *Credit Management*, July 1989.
53. John Donaldson, *Key Issues in Business Ethics*, p.191.
54. Daniel R. Gilbert, Jr., *The Twilight of Corporate Strategy: A Comparative Ethical Critique*, pp.14-15.
55. *NCC (1980) Consumers and Credit, Appendix 1* in Janet Ford, p.32, *The Indebted Society: Credit and Default in the 1980s*; copyright 1988, Routledge, reproduced here by kind permission of Janet Ford and Routledge Publishers.
56. Richard Berthoud, *Credit, Debt and Poverty*, p.16.
57. *Ibid.*, pp.16-17.
58. *Ibid.*, p15.
59. Quoted by Richard Berthoud, *ibid.*
60. Janet Ford, *op. cit.*, p.31.
61. Raymond Lang, 'Analysis of Individuals with Multiple Debt Problems' in Andrew Hartropp, (ed.), *Families in Debt*, pp.19-20.
62. Andrew Hartropp, (ed.), *op.cit.*, p.156: copyright (c) 1988 Jubilee Centre Publications Ltd.: reproduced by kind permission of Jubilee Centre Publications.
63. Richard Berthoud, *op.cit.*, p.12.
64. *Ibid.*, p.15.
65. Office of Fair Trading, *Moneyfax: The Crucial Guide to Credit and Debt*, section 2.
66. Janet Ford, *op.cit.*, p.57.
67. In Ian Cowie, 'Accountants Head the Home Repossessions Table.' p.23, *The Daily Telegraph*, February 9, 1991, © Telegraph Group Limited, London 1991.
68. Quoted *ibid.*
69. M.C. Fleming and J.G. Nellis, 'The Rise and Fall of House Prices : Causes, Consequences and Prospects.' pp.40-41, *National Westminster Bank Quarterly Review*, November, 1990.
70. *Ibid.*
71. *Ibid.*, p.42.
72. Quoted by Ian Cowie in '1000 homes a week are repossessed as mortgage debts rise', p.7, *The Daily Telegraph*, February 15, 1991, © Telegraph Group Limited, London 1991.
73. *Credit Action*, 1990, July to Sept.
74. Remortgage facilities are frequently advised in the press.
75. National Remortgage Corporation leaflet (Undated).
76. Transactions are often by post enabling embarrassing interviews to be avoided.
77. 'A mortgage fix to soothe that repayment headache', p.23, *The Daily Telegraph*, February 9, 1991, © Telegraph Group Limited, London 1991.
78. Barclaycard brochure. Barclays Bank PLC.
79. Member Loan leaflet, FS3977. Automobile Association Financial Services Limited.
80. Abbeyloan from Abbey National leaflet OT4.
81. Choice Personal Finance leaflet, 1, Lyonsdown Road, New Barnet.

82. 'Money Advice Initiative', p.11, *Credit Management*, May 1989.

CHAPTER 4

BANKERS, BUSINESS FINANCE AND CONSUMER LENDING

We have seen that positivism has influenced business operations to the point where technical financial appraisal takes place independently of ethical considerations. As with economics the techniques and tools used mainly address quantitative data as a basis for decision taking. The loan officer is pre-occupied with this detailed analysis of financial information and it is not easy to embed ethical values into it in a meaningful way, but this must be done to secure ethical debt management. If values are not central to lending activities they will be of little significance to bankers. In this chapter banking must speak for itself about lending and as it does, we will relate what is said to our stewardship values.

Stewardship, Banker and Customer

The main players in the retail lending field are banks, building societies, finance companies and credit card companies. Now that the banks provide mortgage finance and the building societies provide current accounts the former distinctions are less important. All financial intermediaries operate within stewardship constraints. In the narrower sense of *entrustedness* we see that resources are entrusted to bankers to control them in a way which preserves their value and gives a return on them. This is illustrated by the sources of funds of a commercial bank, which fall into two main categories. Investors subscribe for share capital and depositors place money in current or deposit accounts. The shareholders entrust the bank with their funds on the understanding that the management will conserve the capital fund, increase its value, and also pay out a cash dividend. Depositors expect the safekeeping of the sums they have deposited, the provision of the services relevant to the class of accounts involved and a modest return by way of interest.[1]

To fulfil its trust the bank utilizes financial *matching* to secure *effectiveness* in controlling the resources committed to it, to secure an appropriate *development* to meet the expectations of those who have entrusted funds, recognizing its *accountability* to all of them. To operationalise entrustedness management must use matching to identify and control value-enhancing operations. The need to maintain sufficient liquidity to meet customer demands for cash is a major constraint. The purchase of gilt-edged securities provides some return with a high degree of safety and liquidity. The more profitable business, however, is lending to customers and here there is significant risk and much lower liquidity. The nature of the stewardship control is specialized and we need to give close attention to the relevant operational considerations to ensure that *fairness* is maintained throughout. Failure in the management of individual accounts results in bad debts and these are contrary to the investors' and depositors' interests. Default also impacts on borrower interests.

The broader sense of *entrustedness* requires bankers to consider their social responsibility and we will identify specific obligations of lending bankers. In the legal context the banker-customer relationship revolves around the debtor-creditor relationship. In the case of the depositor, who entrusts funds to the bank, the latter is the debtor and the customer is the creditor; whereas, in the case of the borrower to whom the bank entrusts funds, the bank is the creditor and the customer is the debtor. The respective duties of banker and customer were summarized in *Joachimson v Swiss Bank Corporation* (1921). E.P. Doyle reports as follows:

> "The banker undertakes:
> (1) To receive his customer's money and cheques... for collection.
> (2) To repay this money ...upon presentation of the customer's written order...
> (3) To give reasonable notice before closing a credit account...
> (4) To maintain secrecy in respect of his customer's account and financial affairs....
> (5) To advise the customer of any forgery of his signature.
> The customer undertakes:
> (1) To seek out the banker if he requires repayment.
> (2) To issue cheques only if he has sufficient balance (or an agreed limit).
> (3) To pay charges as agreed (or reasonable charges)....
> (4) To exercise reasonable care in drawing his cheques...
> (5) To advise the bank of any forgery of his signature."[2]

This summary illustrates that the law regards the parties as having mutual responsibilities requiring each of them to regard more than self-interest. It shows that without acceptance of these responsibilities to one another the relationship would be unworkable. But the expectations here are technical and we will apply our stewardship model to show what ethical expectations we have of lenders regarding customers and society. What is the essence of the banker's role as a lender? Bankers are educated in the skills involved in testing the effectiveness of financial stewardship planning, decision and control of a firm's management in each particular lending case. Similar expertise is used for personal loans.

4.1. Business Finance

4.1.1. PRINCIPLES: EVALUATION OF THE CANONS OF LENDING

The Current Approach to Principles

We begin with the principles of lending. Nicholas Rouse, in his book on lending techniques, reflects upon the difficulties posed for lenders by uncertainty. In considering a philosophy for lending he emphasizes that there is no "magic formula". After gathering all of the relevant information, the lender must apply his skill in judging the degree of risk involved. Rouse states that a methodical approach to appraisal involves five stages of analysis:

> "(1) Introduction of the customer
> (2) The application by the customer
> (3) Review of the application
> (4) Evaluation
> (5) Monitoring and control."[3]

The canons relate mainly to the third stage. Rouse explains that, at the review stage:

> " all the relevant information which is required needs to be tested and other data sought if necessary. Either formally or informally the lender applies what are generally known as the canons of good lending."[4]

He describes the mnemonic CAMPARI, which is used by two major clearing banks. The mnemonic stands for *Character, Ability, Margin, Purpose, Amount, Repayment,* and *Insurance* (security). The main issues Rouse identifies are summarized below.

Character. Reliability of the customer's word regarding data supplied and promise of repayment. Evidence of exaggerated claims. The customer's borrowing record to date. Reason for approach if a new customer.[5]

Ability. Spread of skill and experience among the management team. Any relevant professional qualifications held by them. Commitment to company success. Experience in the area to which funds are to be committed.[6]

Margin. The interest, commission and other fees to be paid. (The interest margin reflects the risk involved and the amount and complexity of the work determine the commission and other fees.)[7]

Purpose. The purpose of the borrowing must be acceptable to the lender. Rouse remarks:

> "Perhaps the facility would not be in the customer's best interests. Customers do tend to overlook problems in their optimism and, if the bank can bring a degree of realism to the proposition at the outset, it may be more beneficial to the customer than agreeing to the requested advance."[8]

Amount. Neither too much nor too little. Proportionate to the firm's resources.

Repayment. There should be a clear source of repayment and an estimate of the degree of certainty that the promised funds will be received.

Insurance. The canons must be satisfied irrespective of available security and the provider of security should understand the consequences of charging it to the bank.[9)]

Stewardship Evaluation

These canons of lending, though technical in emphasis, underline the strengths of our stewardship model. The lender-borrower relationship is one of *entrustedness* and, within such relationships, character is of the utmost importance. The bank is entrusting to the borrower funds that shareholders and depositors have entrusted to it. The questions the banker asks about the borrower have to do with the moral reliability of the customer. The banker wants to be sure that borrowers will be true to their word when promising to repay the funds entrusted. This involves honesty, loyalty and faithfulness but also realism. The borrower must not make rash promises but consider the responsibilities being undertaken and make a commitment accordingly. A satisfactory past repayment record would suggest that the customer takes obligations seriously but it will be important that customers have a real sense of entrustedness. Repayment involves more than willingness to repay. There must be a sufficient level of stewardship competence to ensure that funds will be available for repayment. Entrustedness is made operational through *matching*. Bankers must go further than balancing their own assets and liabilities. They cannot ignore borrower ability to match cash inflows and outflows. Borrower *effectiveness* here is fundamental to repayment. When bankers ask questions about customer ability, they are seeking to form accurate judgements about operational effectiveness in the stewardship tasks of planning, decision and control. So that the bank does not fail investors and depositors in the stewardship of their funds, it must be satisfied that the management of the borrowing firm can use the funds in some *development* that will produce a surplus to meet repayment. The remaining canons are directed to ensuring this stewardship competence, but with specific focus upon the loan proposition. The bank tests the realism of the customer's planned development by considering the purpose of the loan, the amount required, the repayment schedule and the insurance or security for the borrowing. Essentially the banker must test whether or not the proposition would involve good stewardship of the funds to be entrusted to the borrower. In the case of business finance the bank must satisfy itself that the development project would generate sufficient funds to service the loan and to effect repayment. The bank management's stewardship of funds depends upon funds being directed by borrowing firms into successful development projects. It is the surpluses generated by these developments that guarantee the safety of shareholder investments and depositor balances and the returns upon them. Funds not utilized cannot generate returns and the banker is a true intermediary with respect to funds travelling into remunerative development in someone else's business. The banker must thus become an appraiser of borrower stewardship planning. Borrower *accountability* would be empty if the bank itself was lax at this point. *Fairness* to shareholders, depositors and borrowers necessitates the rigorous loan appraisal that ensures that default is minimized. The canons support our view that technical matching must be embedded in stewardship values to maintain ethical lending [10], and the above argument shows that in principle, at least, stewardship values are compatible with the canons of lending. What will be the case with the practice?

4.1.2. PRACTICE: EVALUATION OF LENDING PROCEDURES

To consider the practical application of the canons of lending the existing procedure will be outlined and considered in relation to stewardship values. This will provide a basis for presenting new indicators for operational lending based upon such values.

Existing Lending Procedure in relation to Stewardship Values

The Provision of Information. The starting point for lending appraisal is the information provided by the intending borrower. Customer-generated information however presents difficulties. Small businesses often lack systems to provide quality management information. Management although technically capable in trade terms, might lack detailed accounting knowledge or financial expertise, and be as much in need of advice from their bank manager as a personal borrower. This fact was identified by the *Bolton Report*, which addressed the structure and financial needs of small firms. The report concluded that small businesses could improve their performance.[11] It drew attention to the lack of knowledge of appropriate sources of finance [12] for development and working capital [13], the lack of skill in presenting a financial case to lenders and inadequacies in connection with cost control, credit control and stock control.[14] The Report suggested that banks could take a lead in improving the efficiency of small firms by imposing

> “as a normal condition for the grant of an overdraft to a firm, the production of regular cash-flow statements, or the employment of adequate estimating and budgetary systems ....”[15]

The committee looked to banks to reinforce practices to improve customer efficiency.

Such requirement for information, especially with respect to the future, would reinforce stewardship, as the management would have to think through where they are going and set down a coherent business plan. The *Bolton Report* identified inadequate preparation of a financial case as one of the weaknesses of small businesses.

> "Small firms ... are unaware of the advantages of different methods of raising capital. They are also unskilled in presenting a financial case to potential investors and lenders."[16]

Banks have moved a long way since 1971 in providing information packs to their customers so that the small business can provide the basic information that the banks want. The information pack of the *National Westminster's Small Business Service* contains a handbook entitled *Start Up and Go with NatWest* dealing with such matters as self-assessment, marketing, bookkeeping, and employment. It provides a self-assessment checklist for the assessment stage but it is the second handbook, *Profit by Planning with NatWest* which is the financial guide. Stewardship *entrustedness* must be brought into operations through *matching* and this handbook gives informative guidance concerning business plans, operating budgets, cashflow forecasts and monitoring performance by comparing projected and actual costs. An accompanying booklet provides forms to complete a profile of the business by way of a business plan, a monthly operating budget and a monthly cashflow forecast. If borrowing is required, a booklet entitled *Presenting Your Case: Help your NatWest Manager help you* gives a checklist of the information requirements concerning the business, the purpose of the loan, the firm's market, the firm's profit, the amount required, repayment and security. We conclude that regarding technical matching this example shows the banks have provided helpful customer advice.

The Customer Interview. The primary purpose of the interview is “scrutiny”. Bankers act as financial advisers concerning loans but primarily the interview is about the viability of the lending proposal. This underlines the importance of *effectiveness* by examining whether the borrower can carry the proposed project through to a successful conclusion. The bank interest is in the customer's past record, present position and future plans and this involves independent review of the position audit, objectives and adopted strategy of the firm. Relevant financial statements are as follows: (a) Trading and profit and loss accounts for the previous four years; (b) Balance sheets for the previous four years; (c) Funds flow statements for the periods between the Balance sheets; (d) Profit forecast for the following twelve months, and (e) Cash budget statement for the following period. Items (a) and (b) and possibly (c) are normally prepared by a professional accountant, whereas items (d) and (e) require considerable input from the firm's management. This requires them to quantify their *development* plans by way of financial matching. The fact that these forecasts will be scrutinized by experienced advance control staff should encourage careful thought about resource utilization.

The interview provides an opportunity to assess character, business ability, product knowledge, marketing and manpower planning skills. The bank manager will want to fill any gaps in the information provided and test out the intending borrower as to how unforeseen contingencies would be dealt with. The canons of lending will be in mind. The interview provides the data necessary to proceed to the next stage of assessment. It provides an opportunity for borrowers to reconsider their expected effectiveness in pursuing the proposed development.

Advance Application Form. This form summarizes the information provided by the customer, the amounts and ratios calculated for the purposes of analysis, the relevant information gleaned during the interview and the manager's comments and recommendation. Appended financial statements include: (a) Trading and profit and loss accounts, (b) Balance sheets, (c) Source and application of funds, (d) relevant surpluses and ratios, (e) profit forecast and (f) cash forecast. Some impression of the thoroughness of scrutiny is given by the lists of probing questions suggested by Rouse.[17] It is not necessary for us to consider in detail what Rouse has to say about general economic conditions, materials, labour, expenses and the cashflow forecast to appreciate the banker's intention to satisfy himself that the assumptions for the forecasts are sound . Checkley sums up the reason.

> "It is vitally important to remember that a business, no matter how strong its balance sheet, cannot carry a debt burden greater than can be serviced by the income earning capacity of the business. It is for this reason that great emphasis will be placed on understanding the *trading* 'profile' of the business."[18]

We can see at this point the way in which the economic *development* within the firm is crucial to the safety of investor and depositor funds and returns. It is as crucial to them as if they were direct investors in the borrowing firm and the directors were managing the firm in their interest. Lax lending, producing many defaults, would put the bank out of business and the shareholders and depositors of the bank would lose out. This is the extent to which entrustedness travels through the bank to value enhancing development activities in other firms. The borrower's stewardship failure impinges on distant providers of funds. Rouse defines the second stage in the evaluation as:

> "A critical appraisal of what might realistically go wrong - the likelihood of such events occurring and the effect on the bank's position."[19]

The two stages of the evaluation thus address the risks involved. Rouse explains that, once the lending has been made,

> "the risk lies in the way the customer handles any problems which might arise. The lender's evaluation should concentrate on understanding the borrower's risks and assessing the ability of the borrower to deal with them."[20]

The bank can strengthen its position by taking security but as we have already seen the banker should be satisfied that the canons of lending are met independently of the available security. If the latter has to be resorted to, the realization costs might be significant and the realization value might be disappointing. Risks should therefore be carefully weighed.

Monitoring. Accountability leads on to monitoring. Its importance is underlined by the reasons for small business failure. The 1982 report by the accountants Robson Rhodes related to the first fifty failed businesses giving rise to claims under the *Small Business Loan Guarantee Scheme*. It will be useful to consider some of the significant statistics relating to 48 cases.[21] Of the firms, 38% were existing businesses *already in distress*. New and young businesses were well represented. Manufacturing accounted for 54%. The following are particularly significant.

> 70% of the failed businesses took no advice from bankers or accountants.
> 60% of the firms did not prepare management information.
> 87% had no monthly statement of income.
> 96% had no monthly cash flow statement.
> 94% of the firms had an initial capital gearing in excess of 1:1.

These statistics show that the technical matching we have insisted upon within our stewardship model respecting solvency, forecasting, monthly income budgets and monthly cash flows were being systematically neglected by management. External monitoring of the lending, in the interests of a proper accountability must also have been lacking. Such a situation clearly contravened our emphasis on *fairness* to all parties.

Stewardship Values: Indicators for Operational Lending
Business lending practice has been surveyed avoiding technical detail. We have looked at the basic methodology for appraising business propositions, and have found some correspondences with our model. The basic philosophy for lending, as perceived by bankers, is that the lender must apply lending skills to the information supplied, and then make a judgement that the risks involved are not so great that the customer will be unable to repay. There is no guaranteed formula but the banker is guided by the "canons". These are technical considerations and we will now show the superiority of stewardship values.

Entrustedness and Matching. Entrustedness has a narrow application, focused on resources, and a broader application taking in a range of relationships. The former is brought into operations through matching. Bank funds are entrusted to them and banks must remuneratively use them while maintaining liquidity. A bank's responsibility with respect to matching extends beyond the bank's own balance sheet assets and liabilities. The entrusted funds are being entrusted to others and, to maintain the chain of entrustedness, the bank must be satisfied concerning the borrower's trustworthiness and operational matching. The existing methodology is strong in respect of this narrow entrustedness. The assessment of borrower character included honesty, promise keeping and integrity. Current business finance lending procedures stress matching utilizing final accounts, funds flow statements, profit and cash forecasts and relevant ratios. Banks provide information to familiarize customers with such assessment procedures.

The broad sense of entrustedness brings in, not only investor and depositor expectations but also social expectations. The latter have been formally expressed in the past, as when credit control restrictions have limited lending to export purposes. The abolition of direct controls leaves the banks to develop their own sensitivity to the ethical concerns of society including their own investors and customers. There are some examples of this. Banks have shown responsiveness to the sense of entrustedness to the environment. The Cooperative Bank have expressed commitment to a broader range of social concerns appealing to those investors and depositors who do care what purposes their entrusted funds are used for. Where major room for improvement exists is in building in at the lending approval more protection against the social costs of debt. We will consider this in more detail under consumer lending.

Appraising Effectiveness. The heart of the lending banker's role is the responsibility for appraising the stewardship *effectiveness* of borrowers.

Management failure to secure the planned growth makes repayment impossible. Banker entrustedness is thus intimately connected with borrower operational effectiveness. If the borrower fails, shareholder and depositor funds are lost. The banker cannot justify to either stakeholder group laxity in appraising managerial planning and control in the borrowing firm. Failure would be a failure of trusteeship on the banker's part. Our survey of procedures leaves the impression that appraisal will be thorough. The banker uses the interview to ascertain the necessary information to test the viability of the proposition and reviews the firm's position audit, business plan and its financial implications. The application form summarizes relevant information and the results of the financial analysis of the final accounts. The analysis of the financial statements tests, through matching, components that we have identified as crucial to effective stewardship, including: s*olvency* (consideration of net tangible assets surplus and capital gearing ratio); *liquidity* (calculation of liquid surplus, quick asset surplus and current ratio); *financial control* (consideration of stock turnover and average credit given) and *profitability* (measured by net profit margin and gross profit margin).

Also considered will be non-financial factors such as the firm's markets, products, resources and management. The assessment of the latter includes motivation, character, integrity, experience and specialist managerial abilities. Particular consideration is given to: the loan purpose, the amount required, the repayment plan, the profit budgets, cash flow forecast, forecast sales, economic conditions, and the risks involved. We conclude that this form of lending practice involves thorough scrutiny and testing of the stewardship competence of the firm's management. Banks who follow the practice will be responsible in the appraisal of stewardship effectiveness.

Development and Readiness to Decline Assistance. Pressures on bankers to lend include lending targets, competition, customer connections and the influence of accountants and solicitors.[22] Our model requires that *development* exerts the prevailing influence. Quantitatively, if financial disbenefits of the project in the loan application exceed its benefits, there will be no surplus to effect repayment and the loan should be declined. In broader terms, the human costs of borrowing should be considered, whether arising from default or from preventing default. The health pressures on the self-employed serve as an example. If in connection with risk appraisal the banker concludes that the risks are too great then the proposition should be declined even if adequate security is offered, not only to save a bad debt, but also in the customer's interests. The nature of the development is also important. Although we accept that it is impractical for firms to take external social costs into their profit calculation, we have maintained that our broader view of development requires non-financial costs to be considered. With respect to loans for business purposes, investors and depositors may disapprove of the nature of the product (tobacco), environmental impacts (pollution) or the nature of the venture (speculative). Social pressures are moving bankers on some ethical issues and, consequently, our integrated model would benefit decision-takers.

Accountability and Monitoring. Banks must be prepared to accept a monitoring role regarding loaned funds. This necessity was illustrated by reasons for business failures which suggest that ongoing advice from bankers could have helped to balance the lack of management experience and information. The bank's internal records such as the firm's current account, and the firm's accounts, as supplied, can be used to provide an early warning system. Credit monitoring arises out of the bank's trusteeship. It's responsibility to shareholders and depositors for the safety of the loaned funds requires vigilance to ensure matters proceed according to plan. As entrusted funds travel through the bank to another business, accountability has to be maintained without inhibiting entrepreneurial management.[23]

Procedural Fairness. The banker-customer relationship necessitates *fairness* on the part of both the borrower and the lender. Banks rely on customers to provide accurate information and professionalism requires that the data is approached in a sympathetic, but not sentimental, way. Analysis must be rigorous and far seeing regarding possible risks. The banker should be disposed to help the customer but not against better judgement. Being fair to the customer involves resisting influences inimical to objective appraisal. Supportiveness is important and acceptance by banks of the requirement to provide assistance by way of information and financial advice fits well with this. The small firm particularly needs guidance. The requirement to provide information about future plans does reinforce borrower effectiveness. Information packs doubtless have promotional value but they are helpful introductions to financial matching. Bank assistance is not restricted to brochures. At an interview the proposal can be discussed and specific advice on business plans obtained. Although often criticized for being unfair, some credit has to be given for steps taken. Much criticism relates to fairness in handling remedial action. Loan situations do go wrong and the traditional banking disposition has been to support customers through difficult times in a sympathetic way but not at the expense of objectivity. Increased borrowing and increased interest charges necessitate increased profits in order to effect repayment and unrealistic sales targets do not generate cash. If at times the banker supports an ailing borrower against other criteria it will increase customer losses and probably those of the bank. Such situations require a careful balance on the part of the banker between trustee responsibility to fund providers and the principle of fairness and supportiveness. The bank must minimize losses to shareholders and depositors, but also minimize damage to the debtor, by avoiding premature action that might bring about a crisis and lead to the realization of security at unnecessarily low values. The bank must be fair to all parties.

Conclusion. This summary shows that received lending practice satisfies some of our ethical expectations particularly regarding technical matching and its significance for effectiveness but that it could be better if it followed stewardship values. With respect to the broader implications of entrustedness and development some banks have addressed ethical considerations in connection with loan *purposes*. Failures in maintaining borrower accountability show that vigilance is required in changing economic circumstances. Fairness is so crucial that its implications need greater underlining in the details of lending operations. We have referred to pressure exerted by professionals introducing business and pressure may come from customers pressing for quick decisions or threatening to bank elsewhere. Customer connections can be influential here. If poor quality customer-generated information is relied upon without sufficient probing by the banker, and with too many assumptions being made in order to fill out the details, the banker's decision will not rise above the information upon which it is made. The bank's response to crisis situations may also be inadequate. If security margins are comfortable it may not react with sufficient urgency to adverse trends in the borrower's position and allow it to deteriorate further before acting. Rouse's comments are apposite.

> "Lending is a risky business, and it is inevitable that on occasions things will go wrong, so that with hindsight it appears that a bad decision has been taken. No criticism should be levelled at a lender who has made a reasonable decision based on all the known facts at the time. What does warrant criticism is a lending going wrong as a result of a failure to recognise obvious warning signs and to take remedial action..."[24]

We conclude that tools exist for sound technical analysis but that this cannot of itself address the wider ethical issues that are emerging with business lending. Lending officers need a decision support tool that can address these ethical matters and stewardship values are able to do this without marginalising profit. As we have seen, they have superior worth to technical analysis alone because they focus on entrustedness, not only regarding resources, but in the broader sense which includes social concerns regarding the type of development envisaged and the social costs of debt. Our model also requires bankers to appraise customer effectiveness in a context of ethical stewardship and *stewardship* development requires a broader range of benefits and disbenefits to be brought into consideration, than merely accounting profit. The stewardship model also brings fairness to all stakeholders into the operational lending.

4.2 Consumer Lending

Reductions in sovereign lending to foreign governments, together with some decline in business borrowing channelled large sums into domestic personal sector lending. Financial deregulation made access to wholesale funds easier for some financial institutions reducing dependence upon retail deposits to extend more credit. This has coincided with increased competition and lower lending costs arising from technological innovation. This sounds like good news for consumers who have enjoyed greater access to credit.[25] In recent years sovereign lending has led to unmanageable debt and in 1988 a letter to *Banking World* drew the inevitable parallel with consumer debt.

> "The sovereign debt of the 1970s has proved the nightmare of the 1980s for the banks. Will the personal debt of the 1980s turn into a similar nightmare in the 1990s?" The correspondent comments, "It is very likely".[26]

We will have to agree with this if it can be shown that changing lending procedures do not adequately address a stringent appraisal of stewardship. In the case of sovereign lending, undue preoccupation with borrower character and solvency, to the detriment of broader criteria, produced serious consequences when economic and monetary conditions changed and this prompts us to inspect personal lending practice. Methods to appraise personal lending differ. Individual appraisal, similar to that employed for business loans can be distinguished from the newer credit scoring approach. In looking at these we will draw on issues raised in recent *National Consumer Council* research literature.

4.2.1. STEWARDSHIP AND INDIVIDUAL APPRAISAL

Outline of Current Lending Practice

One example of the still current use of individual assessment is Home Loans. *CAMPARI* is relevant. Concerning *character* and *ability* Nicholas Rouse states,

> "The borrower ought to have a track record with the bank or, if not, evidence should be seen of the track record with his or her present bankers in the form of bank statements. ... The borrower should be in

> employment which will supply a stable income."[27]

Commenting on the *margin*, Rouse refers to this form of lending as very remunerative, with the current mortgage rate being charged, costs of establishing the facility being recouped and opportunities being available for cross selling in connection with insurance.[28] The *purpose* in the case of home loans may not be straightforward.

> "The purchase of a house for owner occupation will almost always be acceptable to a lender. More problematical will be loans requested for the purchase of second and holiday homes..."[29]

The *amount* borrowed should cover purchase costs and essential repairs. With respect to *repayment* practice varies and the term may be from 25 to 35 years. Rouse advises:

> "A typical guideline for assessing repayment capacity is three times the gross main income plus once any secondary income, ... Mortgage rates can rise very rapidly during a year and too much emphasis should not be placed on the ability to service borrowing at whatever the current rate is.... If repayment looks marginal, the borrower should be asked to provide a personal budget..."[30]

Concerning *insurance* a first mortgage over the property will be the prime security. Rouse recommends a first mortgage over sufficient life cover to repay the loan on death and comprehensive and index linked property insurance.[31] On monitoring and control he recommends that home loans should be subject to annual review.[32] He advises scrutiny of the borrower's main account to identify difficulties in meeting payments and encourages a prompt interview with the borrower. He continues,

> "Provided the problem appears to be a temporary one, some restructuring of the repayment programme may be possible. But if there appears to be a permanent difficulty, then the borrower needs to be encouraged to recognise this and face the necessity of selling the property before mounting interest eats into the equity."[33]

Positive Aspects of Current Practice

The impression gained is that technical matching would be well covered. Implicit throughout is the importance of customer solvency and ensuring that the value of the lending (the borrower's new liability) will not exceed the value of the security (the borrower's new asset). There should be no deterioration in the borrower's solvency immediately after borrowing. Maintenance of the solvency margin to provide for unexpected changes is explicitly recognized by Rouse concerning borrowers in difficulty. He recognizes the banker's duty to step in to encourage the customer to sell before the solvency margin disappears. Borrower solvency should normally increase as a result of property investment. The importance of keeping a surplus of income over expenditure is recognized by Rouse and the banker who discovers from the current account that repayment difficulties are being experienced should stress the need to avoid a deficit and offer help in resolving the problem. The bank practice of limiting lending to three times the gross main income plus once any secondary income places some constraint upon the borrower in connection with realistically matching personal resources and expectations. The borrower's objective to acquire an investment product which will (usually) yield capital gains is less problematic than an objective such as immediate consumption. Property purchase is a case where the borrowing strategy has proved successful for many, whereas the saving strategy might not have succeeded because of rising property values. Rouse, as we have seen, regards property loans as very remunerative from the banker's point of view. This implies that they are expensive from the borrower's standpoint. However he does draw attention to "competitive pressures"[34] which serve to moderate the costs. The borrower should be aware that the arrangement involves the bank as creditor having claims against the house. The security formalities should bring this home. The reality of future uncertainties should register with the borrower when the bank insists on insurance. Rouse places emphasis on adequate monitoring and prompt remedial action. In all these respects the received practice for home loans fits well with our technical matching requirements.

Improving Current Practice

Entrustedness and Matching. The complexity of entrustedness means that it cannot be simplistically measured. So, for example, avoidance of insolvency and maintenance of an adequate track record might be accounted for by the fact that such a large financial commitment has not been taken on by the borrower before. It might not indicate capacity for financial stewardship at higher levels as the unreliability of even professionals has shown. In such cases the formalities are not likely to bring home to the borrower the need for realistic economic matching. We saw with business loans that a requirement for profit and cash budgets compels a certain level of stewardship matching which, in itself, is likely to have some useful effect. For home loans, Rouse recommends a personal income and expenditure budget only in cases where "repayment looks marginal"; and applying income guidelines for assessing repayment capacity does not impress on the borrower the necessity of realistic budgeting. It appears that it is particularly the area of realistic financial planning within a context of uncertainty that requires greater attention *before* borrowers commit themselves.

Effectiveness. With home loans effectiveness must be maintained over long periods. Guidelines based on multiples of income place too little emphasis on future relationships between income and expenditure and the crisis arising when the former is falling and the latter rising. Where such a scenario is accompanied by falling house prices and high interest rates it is borrower effectiveness that will decide the outcome. The bank might get out of difficulties by realizing the security, but the borrower loses a home. Lenders need to do more to make their customers aware of the need for long-term effectiveness through ongoing stewardship matching. Some borrowers have too little recognition of the risks involved and are unprepared for crisis. The research literature identifies such things as illness, unemployment and marital breakdown as causes of debt difficulties but budgeting with adequate margins would help.

Development. Home loans satisfy the development objective. From the lenders viewpoint a social need is satisfied. From the borrowers' viewpoint they secure an investment and accommodation. This must not encourage complacency. Many difficulties arise because borrowers aim at the maximum investment and borrowing to secure maximum returns. Our stewardship model does not isolate one value but applies each to each case. The effect of this would not necessarily be that lending could not be approved but that purchasers revise their expectations downwards and borrow less more safely.

Accountability. Annual review and periodic scrutiny of the current account while sound advice will in some cases prove inadequate. Credit availability by a variety of financial products provides the opportunity for borrowers to get into considerable debt before this is reflected in the current account. Borrowers use plastic cards to fend off difficulties only to find that they have aggravated their debt problems. Thus, although bank monitoring and control procedures serve their purpose to some extent, they may not provide a sufficiently early warning where borrowers are running up multiple debts, and our model would look for more sophisticated accountability mechanisms

Fairness. One example must suffice. Because the bank system addresses default, the need for personal monitoring and control is forcibly brought to the attention of the borrower only when difficulties have already been encountered. The bank's support is essentially remedial, not preventative and because of the present opportunities for a very rapid accumulation of debts it is timely for banks to reconsider their emphasis in this connection. Fairness requires that marketing be moderated by customer care involving more attention to whether or not all the different demands of the stewardship model have been met. This would involve more rigorous analysis beyond technical matching.

4.2.2. STEWARDSHIP AND SYSTEMS APPRAISAL METHODS

Outline of Current Practice

Most personal lending is now assessed on a systems basis and this has become a cause of concern.

> "The NPI estimates that at least four million people are already denied access to conventional forms of credit because they score too low on the centralised credit scoring scales used by the big financial institutions."[35]

We will show that the problem is not just one of exclusion and that systems appraisal can be detrimental to those who are able to borrow too. Technological developments, enabling huge databanks, have made it possible to process more loan applications in a cost-effective way. The methods used are *Credit Scoring* and *Credit Referencing*. The former is a statistical method of assessment which awards a score for each piece of information on the score card, such as years in present employment, marital status, time at current residence, age of most recent bad debt and so on.[36] Individual scores when totalled provide the overall score which is compared with a predetermined "passmark". Credit referencing can be used to supplement the credit scoring for all applicants or for marginal cases. Credit referencing agencies hold extensive databases and can provide minimal checks or detailed references. The *National Consumer Council* defines *Credit Scoring* as:

> "a rapid statistical system for (a) assessing the probability that an applicant for a loan will repay the money owed and (b) setting the level of risk the creditor is currently prepared to tolerate in its own lending decisions."[37]

Benefits to the lender include (a) increased approvals, (b) reduction in bad debts, (c) reduced investigation costs, (d) an improved lending portfolio, (e) management information regarding market changes and (f) tailoring to customer groups and risk levels.[38] The *NCC* identifies three main objectives for lenders of scoring systems,

> "to improve the targeting, timing and effectiveness of their default and collection procedures;
> to identify accounts that seem to be running into difficulties and set controls on these customers' credit limits; and
> to identify healthy accounts, and maximise their profitability by targeting more promotions and services at these customers."[39]

Credit Referencing is handled by agencies licensed by the Office of Fair

Trading. The agencies hold data in three categories: identity check, county court judgement and full credit reference. Intending lenders can verify the identity of the applicant, or, in addition, search for county court judgements recorded against the applicant's address or obtain a full credit reference. The information helps check details on applications or can be used to calculate a reliable credit score.[40]

Problems with Current Practice

Appraisal by this method is only as strong as the information provided. There are concerns in two directions. There is the problem of dishonest borrowers. A letter to *Banking World* explains how professional bad debtors can play the system by giving answers that earn the points. Applicants are trusted because in the traditional banker/customer relationship each trusted the other.[41] A second concern is the incompleteness of data held by credit reference agencies. The *NCC* draws attention to the absence of information about utility debts and mortgage and rent arrears, pointing out that these items are often the warning signs of financial distress. Including such items would shape the system into more of a preventative tool than at present.[42] The problem of including this additional information is however recognized by the Council. Its report *Credit and Debt* (1990) lists the additional data to be added, a sample of which includes:

> 46,370 mortgage borrowers in arrears more than six months...; 78,569 electricity disconnections; 61,769 gas disconnections; almost half the local authority rent accounts in inner city and metropolitan areas; 270,000 people in arrears with housing associations ; 7,000 water disconnections; 1,212,000 rates summonses.[43]

The argument for inclusion of the information is stated as follows:

> "Additional data would help to pinpoint more accurately those already struggling to meet household commitments and for whom more credit would bring worse financial problems. It would give a fuller picture of the capacity of an applicant to cope with a new loan, in terms if their income/debt ratio..."[44]

The non-inclusion of default on secured lending is seen to be an important omission at a time of rapid growth in this type of borrowing.[45] Mention is made of a register for consumer credit agreements which would help to identify cases of multiple debt. The report concludes,

> "We believe that the information now acquired through loan application forms, plus the existing ... information on default, are adequate, if not perfect, for credit scoring."[46]

The report recognizes that available evidence on debt and default does not show as high a growth rate as the rise in consumer credit outstanding and expresses the belief that the way in which the consumer data is used in appraisal has contributed to this. However the beneficiaries of this have been lenders and individual borrowers still face overcommitment problems. The commercial fact of life is that controlled arrears are profitable to lenders whereas fast repayment is less remunerative. This makes "stable but slow repayers" attractive to the credit industry and the report maintains that this is reflected in the credit scoring systems.

> "Initial credit scoring concentrates chiefly on the probability of repayment. Evidence of *stable* repayment is highly rated by lenders, whatever the total length or amount. A person's *total* commitment is not necessarily treated as an important variable. Nor is the credit *limit* that is tied to their credit cards."[47]

This emphasis on "financial regularity" rather than "financial capacity" can be disastrous for individual borrowers.[48]

What should we conclude concerning risk assessment systems based on credit referencing and credit scoring? The banker's interest is in their contribution to cost effectiveness, risk minimization, effective credit control and, by the identification of healthy accounts, marketing additional products by specific targeting of advertising. Credit referencing agencies can provide identity checks, county court judgements and bankruptcies and details of current and recent consumer credit agreements. Rapid access to such data enables lending decisions to be made more quickly. From the point of view of those customers who subsequently become casualties we can conclude that the system does not serve their interests. Borrowers who play the system are not real casualties but abuse the system and criminal proceedings would be appropriate, but, there are borrowers for whom the incompleteness of the information collected and the nature of its processing will prove a disaster, because it will provide no brake upon unrealistic expectations on their part and provide no corrective to their lack of financial sophistication. The exclusion of information in connection with default relating to utilities, mortgages and rent means that already struggling debtors are not identified and their situation could be aggravated by further credit. Because stable but slow repayers are profitable business, total commitment is not regarded as an important factor, but the emphasis upon financial regularity rather than financial capacity is not what we are looking for.

Assessment Based upon Stewardship Values

We have based debt management on a stewardship model and expect bankers to reinforce stewardship requirements in their procedures but the systems approach does not reinforce these. The emphasis is on probability and this is the probability of repayment not continuing solvency. Stability of repayment, even of small amounts over long periods, rates higher than total commitment.[49] This is fundamentally at variance with our personal stewardship model for debt management. Linked with this is the need for matching in connection with income and expenditure, yet the *NCC* report states,

> "Lenders tell us that they do not make use of income details because they are not a predictor of the probability of repayment..."[50]

Lenders thus disregard the fundamental of individual ability to repay knowing that their own position is covered by the probability, built into the credit scoring system, that for every bad payer there will be three or four good accounts depending upon where the cut off score is set. This is good for profits but not for some borrowers. Adjustments are needed in risk assessment systems to take capacity to repay into consideration.[51]

There are also the concerns that systems approaches make spending large amounts too easy [52] and that "indebtedness is sometimes made easy by businesses that provide credit."[53] This illustrates that the systems approach fails to protect borrowers. Were borrowers following our stewardship model easy credit would be less of a problem, but the fact that they rely on lender assessments leaves them vulnerable. They lack the financial acumen of business managers and in times of easy credit at high interest rates they can quickly fall prey to persuasive advertising of a manipulative nature. The *NCC* has predicted a rise in debt difficulties as the number of people taking out second mortgages in order to pay for luxuries increases and wants new safeguards to discourage lenders from making loans to those already in debt.[54] Lady Wilcox, of the *NCC* has said that they wanted,

> "to prevent over-enthusiastic use of credit by undiscriminating and poorly informed consumers."[55]

A systems approach provides little support for the customer in connection with basic stewardship matching. We incorporated in our personal debt management model the importance of realistically matching expectations with resources, but the systems approach does not require consumers to do this, although it assumes that they will have. With respect to distinguishing objectives such as immediate consumption, durables and investment, the systems approach is largely indifferent. The stewardship approach stresses the importance of choosing between strategies for purchasing. The savings strategy is realistic for certain types of purchase but the easy credit makes waiting unnecessary and the more insistently credit is promoted the greater the danger of irresponsible buying involving onerous credit terms. This is especially so when marketing literature emphasizes the advantages of "credit" but gives a low profile to the burdens of "debt".

In our stewardship model we allowed for uncertainty and borrower risk.

These systems provide for lender risk by accepting specified break-even odds but this provides no protection for the *particular* borrowers who become the casualties allowed for in the aggregate. The system cannot predict *who* will be affected by common causes of default such as sudden illness, unemployment or marital breakdown.[56]

Our stewardship model advocates starting out with realistic budgets, continuing regular monitoring of actual against budgeted expenditure and taking decisive remedial action. Performance scoring systems enable lenders to distinguish good and bad accounts and to improve the efficiency of their default procedures, but the emphasis upon regularity of payment is capable of masking developing overcommitment until it is serious. It provides no early warning system to reinforce good stewardship through effective domestic budgeting. Thus, on balance, we have more cause for concern in connection with a systems approach to risk assessment than with individual appraisal using more comprehensive criteria. This latter method, it would appear, will serve the borrower better by focusing attention upon activities inherent in good stewardship: considering solvency, solvency margins and surplus income, matching resources and expectations, resisting overcommitment, assessing borrowing risks, maintaining adequate monitoring and so on. It still requires supplementation regarding wider ethical issues as we have seen.

4.2.3. CONCLUSIONS

The Lender's Viewpoint

Before summing up we look at a banker's point of view. Andrew Howe Browne of the *British Bankers Association* puts the banker's case in a feature article in *Banking World*.[57] He accepts that,

> "Outstanding personal credit in the UK rose by about 93 per cent in the four years 1986 to 1989..." but maintains that "stories of individual personal tragedy, of the sort which make news, should not lead directly to the conclusion that this state of affairs is wholly undesirable in political, economic and moral terms."[58]

He makes the following points.

1. The personal sector solvency ratio of £5 assets to £1 borrowed is satisfactory.
2. Lending increases are balanced by an equal increase in net financial worth.
3. There is no explosion in borrowers, existing borrowers are using more credit.
4. The banks underpin the savings strategy by high-rate deposit accounts.
5. The ratio of savings to income rose from 4.9% in 1988 to 7.7% in

1990.
6. 70% of borrowing is mortgages pursuing the investment objective.
7. Much of the rest is for durables including transport for earning a living.
8. Some personal credit is used by the self-employed for commercial purposes.
9. Banks finance according to customers' requirements.
10. Credit cards account for only 2% of personal borrowing.
11. Over 40 per cent of cardholders repay within the interest-free period.
12. Credit cards have an interest-free period. Average borrowing is four months.
13. Economic success depends on the distribution of financial resources.
14. Consumer credit helps industries sell their goods and is not unproductive.
15. Periodic payment of interest is as rational as regular rent for durables.
16. Bankers are not irresponsible lenders but exercise appropriate caution.
17. Bankers do not merely "set their rates at a level which statistically will tolerate an acceptable percentage of bad debts..." They want customer loyalty. The comprehensive financial picture of a customer built up over time helps managers assess the credit that can be accommodated.
18. Banks take care about the financial standing of completely new customers in contrast to retailers who might act on an address, signature and credit card.
19. The problem rate on bank loans in 1987 was half that of finance companies.
20. The rate of growth of default is lower than that of overall personal credit.
21. Bank repossessions are below that experienced by building societies.[59]

These are useful points and reveal some sensitivity to various ethical points to which we have drawn attention, including personal solvency, the savings strategy, investment objectives and awareness of the significance of default and repossession in individual cases. This reveals that bankers could find our comprehensive model attractive to them. We will refer to Browne's points in our summary.

Issues In Personal Sector Lending

Entrustedness and the Erosion of the Borrower-lender Relationship. The banker-customer relationship is endangered by the changing times. Some consumers have become professional bad debtors and can no longer be trusted.

The relationship is also being eroded from the banking industry side because the desire for cost-effectiveness coupled with innovative methods of loan appraisal have reduced personal interaction. More banker contact is at arm's length through documentation. The vacuum is filled by agencies for consumer advice and assistance which borrowers contact when their financial situation is already serious. They must of necessity be remedial, whereas the trust built up in the banker-customer relationship has great potential for preventative action.

Matching and the Decline in Supportiveness. The banker's financial adviser role is endangered by new methods of marketing financial services and appraising loan applications. Direct mail promotes credit in an impersonal way by brochures influenced by marketing objectives. Systems approaches to loan appraisal reduce the personal interaction between bankers and customers regarding major financial decisions and the lack of timely advice could contribute to subsequent default. Limiting mortgage lending to a multiple of borrower income reinforces the need to match expectations to resources but systems appraisal involves an incompleteness of information with regard to total commitment and rules out a rigorous testing of the realism of the borrower's expectations in relationship to resources.

Easy availability of credit can deter consumers from using the savings strategy and precipitate irresponsible buying and it is probable that the borrowers who are least aware of the demands of our stewardship model will be the ones most likely to be adversely influenced by promotional material. The banks' readiness to underpin the savings strategy is welcome and this is socially advantageous because it makes the choice between the borrowing strategy and the savings strategy a realistic option.

Effectiveness and Loan Appraisal. The heart of the banker's role is the appraisal of the personal borrower's stewardship effectiveness and our model for personal debt management includes solvency, income and expenditure and expectations and deals with long range plans within a context of uncertainty involving objectives, strategies and risk. Our cornerstone in financial terms is survival and security based on solvency, but what we have found is *a decline in the appraisal of solvency.* The latter is involved in individual appraisal, whereas the systems approach is much less satisfactory at this point. The emphasis upon stable repayment, even if slow, overshadows the importance of the borrower's total commitment. While reference to the maintenance of a solvency ratio of 20% in the personal sector instills a sense of proportion concerning personal debt, such an averaged figure bears no relevance to the individuals at the extremes and must not become a basis for complacency with respect to the significant number of borrowers who fail to take solvency seriously enough. We conclude that more needs to be done than at present.

Development. Promotional material often encourages consumers to borrow funds to spend on anything the customer desires. A very strong commitment to total family welfare and long range planning for solvency is consequently necessary to balance consumer thinking about choices available. The stewardship values would be helpful in encouraging consumers to consider the broader picture.

Accountability and Inadequate Attention to Domestic Budgeting. If too much emphasis is placed on the past track record as opposed to future domestic budgets the need for the latter may be overlooked by the borrower and the banker might not test any budgets prepared. The past record might be good because the customer has not previously taken on large financial commitments. The application of an income multiple guideline will not bring the borrower into a realistic interaction with the income and expenditure figures. This might leave the borrower unprepared for crisis. This is even more likely with systems-based appraisal where the borrower's planned matching of income and expenditure is not seriously scrutinized. This failure to test the borrower's capacity to repay is not in the best interests of customers even though many of them will not suffer financial crisis in consequence. Some will be casualties of an insufficiently probing system and the lack of an appropriate financial accountability system may result in subsequent default.

Fairness Issues (a) Too Little Attention to the Adverse Impacts of Credit Promotion Materials. The breakdown of bank personal lending shows it in a better light than headlines might suggest. Most goes into objectives involving investment in property, business assets or labour saving or job-related durables. This is not incompatible with good stewardship. There is some strength in the argument that bankers merely respond to their customers' requests for financial assistance and do not shape the secular life-styles that give rise to these. Relationships between consumer wants, credit availability and promotional activity are, however, more complex than this argument implies. Credit availability influences expectations and future plans without sufficiently emphasizing uncertainty.

Fairness Issues (b) Too Little Awareness of Borrower Risks. In the case of property loans security formalities should bring home to the borrower the seriousness of debt and future uncertainties. Although it is important to reckon with uncertainties before entering into commitments, there still appears to be too little recognition of the risks among borrowers, resulting in them being unprepared for crisis situations. Such situations feature prominently as causes of debt difficulties and necessitate adequate solvency and income margins even if these require the borrower to revise expectations downwards. Obviously, even a systems approach cannot predict illness, redundancy or marital breakdown in particular cases, although it can be used to minimize the lender's risk. This is of no advantage to *individual* borrowers who subsequently encounter unforeseen difficulties and more needs to be done to bring home to borrowers the seriousness of financial commitments. If routines for credit approval are too relaxed this will make matters worse.

Fairness Issues (c) Need to Deter Over-enthusiastic Use of Credit. The examples of raising large sums in a short time illustrates the eagerness of stores to approve rather than decline credit requests. The outcome in terms of spending power was impressive but the repayment consequences would have been very serious. In times of easy credit a responsibility falls on lenders to deter over-enthusiastic use of credit arising from the impact of persuasive advertising. They must be prepared to disappoint the financially unsophisticated and undiscriminating customer who is happily prepared to be sucked into the debt trap.

The Superiority of Stewardship Values

S. Carr, *Conservation on farms: conflicting attitudes, social pressures and behaviour,* found that it is over detailed applications that stewardship values, which secure agreement when discussing general issues, are in danger of being overridden. Although farmers accepted the advantages of conservation in general, when decisions about specific farm practices had to be made, business considerations outweighed conservation.[60] This finding supports our embedding of technical matching *within* a stewardship value framework to provide an operational decision model. We have examined the principles and practice of lending for business and consumer borrowing and have considered individual and systems appraisal methods. Although we found the latter much less satisfactory from a stewardship perspective, it is now, at the time of writing, being extended to the appraisal of business finance. This renders our evaluation, using the stewardship concept, all the more important at the present time. In this sub-section the superiority of the stewardship values will be illustrated by reference to business ethics authors.

We derived our stewardship values by analysis of the concept in *Chapter 2* and are applying them as an interrelated set of values. We saw in *Chapter 1* the tendency to select and adopt only those values that match existing practice. Thus merchant capitalists approved of frugality and thrift if it provided them with business funds. More recently, Alicia Barcena and Noel D. Payne, in an article entitled "Stewardship and Sustainable Development", have drawn attention to the same problem of manipulation for one's own ends and caution that:

> "... applying the concept in fragmentary and selective ways to respond to the objectives of different interest groups can result in more, not fewer, imbalances."[61]

We must, therefore, apply our key values as identified in *Chapter 2.*

Entrustedness. We made reference to the bank's entrustedness in relation to investors and depositors. Its management is in a position of trust, entrusted with the funds of others. This introduces the idea of service, with which the banker is quite familiar. But, as Jackall found, managerial service can become service to other managers or company profit.

> "Success in the executive office does have a wider meaning than personal promotion. ... It includes service to the interests of others in *their* promotability; and often this serving involves the muting of criticism..., holding back new ideas ... collaborating in every way with the known views and policies of the current CEO. Service to company profit is also a sine qua non of promotability in the corporate hierarchy."[62]

This presents a bleak picture of managerial success. The latter does not arise out of innovative and creative managerial actions to meet legitimate stakeholder expectations but the complex of internal relationships. Stewardship, by way of contrast, focuses neither upon profit, as an impersonal criterion, nor upon managerial relationships but upon relationships to the varied stakeholders, including investors, depositors and borrowers, demanding service in *their* interests. If social contribution is to have any meaning at all it must involve broader horizons than personal or peer promotion. Dees points out that agency theory's assumption of self-interest is inadequate, because, if people did act upon such an assumption, trust would be undermined.

> "attributing all human behaviour to self-interest is adopting a worst case scenario: A person will always put her own interests ahead of the interest of others whenever she can get away with it. If people act on that assumption, trust is undermined and the worst case scenario could become self-fulfilling."[63]

In contrast to such an assumption, stewardship emphasizes entrustedness and encourages a relationship based on mutual trust, where the principal can expect the agent to act in the principal's interests. Evan and Freeman, discussing action in other person's interests, see the fiduciary relationship as broader than that to shareholders.

> "Specifically we include suppliers, customers, employees, stockholders, and the local community, as well as management in its role as agent for these groups."[64]

We have maintained that stewardship involves entrustedness between depositors and bankers, bankers and borrowers, as well as investors and bankers. It provides moral undergirding for the legal banker-customer relationship and broadens its scope into one of far-reaching mutual responsibilities. This should be the basis of banking practice with respect to depositors *and* borrowers so that the banker's stewardship is not narrowly focused upon return to investors, but embraces customer interests.

We may illustrate by the loan situation where the banker entrusts funds to borrowers for specified periods of time. The bank relies on the customers' characters as honest, loyal and faithful people and trusts that the customers will not intentionally distort financial reports submitted or decline to make repayments. But the customers also trust the banker and, to a greater or lesser extent, depending upon their financial acumen, entrust their personal welfare to the banker by relying on guidance and advice given. Entrustedness is thus two-way and this makes the significance of the other values all the more important.

Matching. Appropriate forms of objective matching are essential for business and domestic stewardship and the banker's trusteeship to shareholders and depositors and supportiveness to borrowers, necessitates thorough appraisal of the borrower's stewardship effectiveness in relation to the total range of measurable factors. These have respect to the present and the future and, as we have seen, include assets and liabilities, resources and expectations and future income and expenditures. Current repayment efficiency is no guarantee with respect to these. As Solomon explains:

> "Efficiency, ... is not the same as *effectiveness,* and the latter, unlike the former, presupposes complex goals and purposes rather than simple measurable functions."[65]

Our model addresses complexity by integrating consumer solvency and costs and benefits into long-range planning for family welfare enhancement, matching consumption and work. Loan appraisal, in consultation with the customer, gives the borrower opportunity to re-think the implications of a borrowing strategy in the light of all the circumstances. This may result in the identification of unforeseen problems and a consequent change of plans, so that the proposition becomes safer from the *borrower's* point of view. Individual appraisal allows, at the outset, for this broader consideration of issues, whereas debt counselling brings a second mind to bear only after problems have arisen and default has occurred. The lender is best placed at the outset to apply appropriate measurements to scrutinize loan propositions in a way which will help to identify potential hazards *to the borrower* and quantify the financial effects of something going wrong.

Effectiveness. Effectiveness in debt management has particular reference to effective repayment. Failure to repay leads to default resulting in bad debts and losses for shareholders and depositors. Effective management thus necessitates operational practices that will ensure effective repayment by borrowers. We saw in *Chapter 1* the problem of lax lending procedures. In analyzing the near collapse of the American savings and loans industry Robert F. Hartley emphasizes the central position of managerial planning and decision making. He recognizes that external factors, such as the 1986 oil price plunge, played a part, and that the US Government were inept in some of their regulatory provisions; but he sees the heart of the matter to be a failure of managerial trusteeship resulting from abuses such as:

> not prudently assessing rewards against risks in proposals,
> operating beyond reasonable means, and
> not keeping a tight rein on costs.[66]

In Texas, part of the problem was irresponsible managerial response to deregulation. *Savings and Loans* lending power had been restricted to the lesser of (a) the purchase price or (b) the appraised value of the project. Deregulation, resulting from overriding federal relaxations, permitted lending to 100 per cent of appraised value, irrespective of actual purchase price, which could be much lower. This provided an opportunity for institutions to lend on much riskier ventures where the asset(s) purchased would provide insufficient security value for the borrowing.[67] The lesson is clear. It is by no means certain that loan industries left to themselves will pursue responsible lending policies, unless there is an acceptance of core values that reinforce stewardship trusteeship *at the operational level.*

Default has disastrous effects for borrowers as businesses are liquidated and homes repossessed. We have therefore concluded, that loan appraisal is fundamental, not only to managerial entrustedness with respect to shareholders and depositors, but also to borrowers. The latter expect supportiveness from their banker to compensate for their own limitations in finance. This is particularly important for small business proprietors and personal customers. This supportiveness is provided by way of information packs and face to face advice. We have identified, however, that the personal aspect is being eroded by increasingly arm's length promotion of services by direct mail and impersonal systems appraisal methods. These methods serve the banks well in identifying customers and unacceptable risks, but they do not serve borrowers to the same extent. Borrowers, in part, entrust their future to the guidance of the more experienced banker and may assume that loan acceptance confirms the viability of their financial circumstances. But, as we have seen, systems appraisal, with its emphasis on regular repayment, leaves aside crucial factors concerning long term financial survival such as solvency, domestic budgeting, planning for future uncertainties and total commitment.

Development. As specified in our stewardship model development embraces economic development and, as we have seen, technical accounting analysis is used to measure this. But we are not limiting development to this but have a moral understanding of it, which includes social and environmental issues. We are not suggesting that these should be reduced to quantitative measurement as in cost-benefit analysis, but our stewardship model allows for their inclusion within the decision process as qualitative factors.

Accountability. In loan transactions, funds belonging to shareholders and depositors are entrusted by the bank to borrowers. The banker is accountable to the providers of the funds and must hold the borrower accountable too. In consequence, monitoring has an important place in debt management. Koford and Penno discuss monitoring in the context of agency theory, making the point that "the value of monitoring of given internal control systems is reduced as the proportion ... of ethical agents increases."[68] Ethical agents can be relied upon to act in the interests of *others*, whereas self-interested economic agents require very close monitoring because they will always take the course of action that will maximise *personal* benefit. This is true both for internal and external relationships. Monitoring is expensive and consumes resources and is to the financial detriment of shareholders and depositors. The latter stakeholders would benefit from lending directed to ethical agents because the reliability of character of borrowers and effectiveness of personal financial management are not only the best guarantees of repayment but also promote cost-effectiveness. The adoption by lenders and borrowers of our stewardship model would reduce internal and external monitoring costs, increase asset value enhancement and benefit all stakeholders.

Fairness. The stewardship values considered above should not be considered in isolation from fairness. The stewardship values are interrelated and require application together. It is thus inappropriate for bankers to set aside all that has been said above by an exclusive focus upon financial entrustedness, as though their stewardship responsibility is discharged by maximizing returns to investors and depositors. As Dees Explains:

> "Many of the ethical questions that can be raised about contractual

> relationships fall under the general heading of 'fairness,' and are based on a common concern to limit the extent to which a stronger party might gain at the expense of a weaker party."[69]

There are many ways in which borrowers can be disadvantaged by possessing less financial expertise than bankers. Credit promotion materials may have adverse impacts on consumer behaviour. Credit appraisal methods may adequately assess the banker's risk but not that of the *individual* borrower. Remedial action may be insensitive or lacking when necessary. It is thus incumbent upon bankers to try to see the ethical "big picture" to be fair in all circumstances. Fairness is regarded by Solomon as a basic business virtue, not so much an ideal as "a basic expectation".[70] He goes on to say,

> "It has to do with honesty, dependability, and trust, insofar as mutual agreement is, in business, the hallmark of fairness. It also has to do with the notion of equivalence or 'equity', the equal value of what is exchanged, whether it be goods, work or wages." He stresses "the importance of mutual agreement, which, writ large, is what we call 'market value'"[71]

We must not interpret this in a mechanistic way. Mutual agreement, equivalence of value and market value, will not, of themselves, secure fairness in debt management. Borrowers may agree to particular arrangements precisely because they are unaware of the potential pitfalls. The focus of credit promotion materials will usually be upon the advantages of having present funds to spend. The loan appraisal may concentrate upon past and present repayment performance. Borrowers may lack the training to identify potential weaknesses in their personal finances. Thus, to be fair, bankers must utilize their greater experience to do as thorough an appraisal in the interests of the customers as they do in the interests of their banks. Our stewardship model facilitates this task by bringing ethical considerations and technical finance together in the appraisal.

Notes to Chapter 4

1. For a recently published full list of sources (1995) see Jeff Madura, *Financial Markets and Institutions,* pp.421-427.
2. E.P. Doyle, *Law Relating to Banking*, pp.A22-A23.
3. C. Nicholas Rouse, *Bankers' Lending Techniques*, p.4.
4. *Ibid.*, p.5.
5. *Ibid.*, p.6.
6. *Ibid.*
7. *Ibid.*, pp.6-7.
8. *Ibid.*, p.7.

9. *Ibid.*
10. Rouse recognizes that professional accountants and solicitors are an important source of new business and that the lender must not succumb to pressure they might seek to exert. *Ibid.*, p.5.
11. J.E. Bolton (Chairman), *Small Firms: Report of the Committee of Inquiry on Small Firms*, p.14.
12. *Ibid.*, pp.120,150.
13. Banks are a major source of short-term finance including seasonal needs. For recent comment (1995) see Moyer, R. Charles et al., *Financial Management,* p.699.
14. J.E. Bolton (Chairman), *op.cit.,* p.113.
15. *Ibid.*, p.119.
16. *Ibid.*, p.113.
17. C. Nicholas Rouse, *op.cit.*, p.77-78.
18. Keith Checkley, *Finance for Business: A Guide for the Lending Banker*, p.69.
19. C. Nicholas C. Rouse, *op.cit.*, p.8.
20. *Ibid.*
21. See Keith Checkly, *op.cit.*, pp.12-19.
22. Refusal to lend could lead to a complaint to the Ombudsman if maladministration is alleged. Office of the Banking Ombudsman, *The Banking Ombudsman Scheme, Annual Report, 1996-97,* p.20.
23. One of the problems of loans to foreign companies is the difficulty of monitoring the lending. See Jeff Madura, *International Financial Management,* p.650.
24. C. Nicholas Rouse, *op.cit.*, p.106. On lending supervision see L.S. Dyer, *A Practical Approach to Bank Lending,* pp. 43-51.
25. Interest rates have remained high however for personal borrowing.
26. K. Sunderland, 'Sovereign Debt - Sweet Dreams or Nightmares?", p.4, *Banking World*, December 1988.
27. C. Nicholas Rouse, *op.cit.*, p.20.
28. *Ibid.*, pp.20-21.
29. *Ibid.*, p.21.
30. *Ibid.*, p.21-22.
31. *Ibid.*, p.22
32. *Ibid.*
33. *Ibid.*
34. *Ibid.*, p.20.
35. Lea Paterson, 'Low Income Groups Face Credit Exclusion', p.18, *The Independent,* 17th November 1997, reproduced here by kind permission of *The Independent.*
36. National Consumer Council, *Credit and Debt: The Consumer Interest*, p. 157.
37. *Ibid.*, p.62.
38. *Ibid.*
39. *Ibid.*, p.63.
40. *Ibid.*, p.61.
41. K. Sunderland, *ibid.*

42. National Consumer Council, *op.cit.*, p.64.
43. *Ibid.*, p.65.
44. *Ibid.*, p.64.
45. *Ibid.*, p.65.
46. *Ibid.*, p.66.
47. *Ibid.*, p.67.
48. *Ibid.*, p.67-68.
49. *Ibid.*, p.67.
50. *Ibid.*, p.68.
51. *Ibid.*
52. See Tom Sorell and John Hendry, *Business Ethics*, p.68 for an example.
53. *Ibid.*
54. Outstanding consumer credit can double in a five-year period.
55. National Consumer Council, *op.cit.*, p.v.
56. *Ibid.*, p.63.
57. Andrew Howe Browne, 'The Banks and Personal Credit', pp.17-19, *Banking World*, January 1991.
58. *Ibid.*, p.17.
59. *Ibid.*, pp.17-19.
60. S. Carr, *Conservation on Farms: Conflicting Attitudes, Social Pressures and Behaviour,* Abstract.
61. Alicia Barcena and Noel D. Payne, 'Stewardship and Sustainable Development', p.34, *National Forum,* vol. 75 no. 1, 1995.
62. James W. Kuhn and Donald W. Shriver, Jr., *Beyond Success: Corporations and their Critics in the 1990s,* p.264.
63. Norman E. Bowie and R. Edward Freeman (eds.) *Ethics and Agency Theory: An Introduction*, pp.9-10.
64. William M. Evan and R. Edward Freeman, 'A Stakeholder Theory of the Modern Corporation: Kantian Capitalism' in Tom L. Beauchamp and Norman E. Bowie (eds.) *Ethical Theory and Business, Fourth Edition*, p.76.
65. Robert C. Solomon, *Ethics and Excellence: Cooperation and Integrity in Business*, p.30.
66. Robert F. Hartley, *Business Ethics: Violations of the Public Trust,* p.196.
67. *Ibid.*, p.190.
68. Kenneth Koford and Mark Penno, 'Accounting, Principal-Agent Theory, and Self-Interested Behavior, in Norman E. Bowie and R. Edward Freeman (eds.), *op.cit.*, p.132.
69. J. Gregory Dees, 'Principals, Agents and Ethics' in Norman E. Bowie and R. Edward Freeman (eds.), *op.cit.*, p.41.
70. Robert C. Solomon, *op.cit.*, p.209.
71. Ibid.

CHAPTER 5

MORTGAGE FINANCE AND MONEY ADVICE

5.1. The Issue of Justice

Economic activity takes place in different socio-economic systems. The particular structure adopted determines who enjoys the social benefits and who bears the burdens. *Distributive justice* is concerned with the right distribution of the benefits and burdens among the members of the community. In order to determine what is just suitable criteria are required. Beauchamp and Bowie explain that "material principles of justice" are used to "put material content into a theory of justice".

> "Each material principle of justice identifies a relevant property on the basis of which burdens and benefits should be distributed."[1]

They itemize the major principles of distributive justice as equal share, individual need, personal rights, individual effort, societal contribution and merit. We must not assume that any one of these suffices on its own to secure distributive justice. Michael Walzer cautions against the expectation that any single distributive criterion will be sufficient in our complex social circumstances. E. Clinton Gardner states:

> "While distributive justice may be defined at the most abstract level in terms of one basic universal principle such as equality (desert, need, social utility, ability, legal entitlement), other moral criteria are also needed in the application of this norm to different kinds of goods in particular situations"[2]

Walzer draws attention to the importance of the family and kinship in connection with distributions according to various conventions and affections.

> "Important distributions are carried out within the family and through the alliance of families. Dowries, gifts, inheritances, alimony, mutual aid of many different kinds: all these are subject to customs and rules that are conventional in character..."[3]

We have already drawn attention to the importance of family considerations in our discussion in *Chapter 3*, and we must now relate principles of distributive justice to domestic owner-occupation.

Distributive Justice and Rights

Home ownership raises the question of *rights*. In this context the definition in mind is the right to hold, occupy and use without interference from others. Interest in owner-occupation requires some stability regarding property rights. Robert Nozick seeks to establish such stability on the basis of *acquisition* and *transfer*. He formulates his view as follows.

> "1. A person who acquires a holding in accordance with the principle of justice in acquisition is entitled to that holding.
> 2. A person who acquires a holding in accordance with the principle of justice in transfer, from someone else entitled to the holding, is entitled to the holding.
> 3. No one is entitled to a holding except by (repeated) applications of 1 and 2." [4]

Such principles underpin the present operation of the housing market.

Private ownership is closely related to family stewardship. The scope for the latter is obviously greater, with respect to accommodation, in the case of owner-occupation as property development shows. In the case of home ownership property extension might be a viable alternative to moving to new accommodation in the event of increased family size. This might limit stress and conserve resources in the long term and thus be more beneficial to the family. Such an alternative will not normally be an option in rented accommodation. Another example would be the limitation of stewardship with regard to providing for one's dependants. The equity in an owner-occupied home is a welcome financial provision in the event of premature death. In other circumstances it is one means of providing for one's posterity and the kind of distribution to which Walzer refers would result in financial benefit to the children.

Needs, Equality and Distributive Justice

This is not to suggest that everyone, irrespective of other considerations, should be accorded entitlement to owner occupation. To seek to maintain such a distribution on the grounds of *need* would be fraught with problems. It cannot be doubted that accommodation is a basic need, but this does not require ownership. Accommodation can be made available through rented property, sheltered accommodation, private boarding houses and so on. Not everyone wants to undertake the financial commitment involved in the acquisition of property or is in a position to do so. But this need not prevent appropriate accommodation being provided. One of the strengths of the voluntary sector has been its capacity to provide shelter even for those who are financially destitute. Beauchamp and Bowie, explain how a two-tiered system of access to goods and services can work providing for social coverage for basic needs alongside private purchase of other goods and services. The first tier provides for distribution based on need but better goods and services can be provided through "an economic system on the second tier."[5]

When the principle of *equality* is introduced, it might be argued that everyone should have an equal opportunity of home ownership and we need to consider whether this can be sustained. There is an issue of stewardship here. Expensive asset acquisition necessitates foregoing immediate consumption with a view to longer-term benefits. The time period involved in accumulating sufficient capital to purchase domestic property is too long to be practical; but the availability of mortgages enables those on modest incomes to acquire their own home.

Individual Effort and Equality of Access

Such a case could be argued if the distinguishing consideration related to *individual effort.* Some make the effort to save with the relevant planning, control and sacrifice involved. It would be a strange morality that would require that others, who did not care to plan ahead or save, should nevertheless experience the same asset enhancement, even though they had voluntarily chosen to use their income for immediate benefits. Asian societies have a propensity to save and there are always individuals who want to use income for long-term investment whereas others prefer to use their income to finance holidays or other short-term objectives. This choice may be perfectly legitimate under the circumstances, but in some cases it may arise from an unreadiness to allow family security to impinge upon current life-style.

It is one thing to argue that personal decision results in inequalities of personal wealth in cases where personal income is the same, but additional problems arise where inequalities of income result in home ownership being viable for some earners but not for others. Even here, however, where the case for intervention seems obvious, difficulties of application arise. If assistance were granted to those on low incomes in order to assist with house purchase, the application of a principle of equality of access would be problematic. We may compare two different low-income families, *Family A* with two parents and one adult child and *Family B* with two parents and three adult children. We assume that all qualify as low-income earners. *Family A*, with three incomes, might need more support than *Family B*, but, if the latter thought they needed four separate houses, would the support be allocated per family or per individual? The public sense of fairness may stretch to supporting *Family A* to a greater extent than *Family B* but not to supporting *Family B* to purchase four separate homes. If the allocation was made per family, individuals in *Family B* might claim that it did not represent equality of access for them. This dilemma underlines the importance of links to personal responsibility, contribution and stewardship in connection with *asset ownership*, whereas there might not generally be the same insistence with respect to *accommodation* of one sort or another.

Stewardship as a Benchmark

The uniqueness of each property presents a problem with respect to domestic accommodation. Even properties built to exactly the same design have a unique location. This factor makes valuations arbitrary and surveyor's valuations reflect the outcomes of recent exchanges of properties of similar design and location. The problem is compounded by owner redevelopment. The owner-occupied home provides scope for personal creativity and is important for the domestic culture. Older public housing stock of various kinds resulted in standardized and regimented environments that limited individual creativity and appreciation of property, particularly among the younger generation. The sale of public housing stock released an interest in property that has led to imaginative renovations to individual taste that have upgraded local built environments.[6] Regarding basic principles of distribution and housing it is thus clear that we do not even have *two* identical units of the commodity involved and this cautions us against grand schemes to secure distributive justice in this case. Stewardship does, however, have something to offer here in relation to distributive justice. The concept provides a benchmark to keep in balance the social and resource perspectives involved in distributive justice so that these can be applied in particular cases.

Stewardship and the Basic Economic Activity Model

Michael Walzer emphasizes the importance of recognizing the complexity involved when considering human society as a distributive community. Although this is but one aspect of human society it involves more than sharing finished products. We come together not only for purposes of exchange, but also to produce the goods and services that enter into the exchange. Work itself is distributed according to the division of labour and this affects individual standing and reputation as well as wealth. Walzer thus affirms:

> "The idea of distributive justice has as much to do with being and doing as with having, as much to do with production as with consumption, as much to do with identity and status as with land, capital or personal possessions."[7]

As we have seen in *Chapter 2* stewardship is embedded in the fabric of life. We will consider stewardship in the context of economic activity to show how it relates to criteria for distributive justice and then apply it to mortgage finance to show that it does facilitate multi-faceted analysis of complex cases.

The basic economic model is as follows. The community of human beings has a multiplicity of needs. In order to meet these needs members of the community must engage in work. Those who do so for economic reward are called the labour force. As we saw in *Chapter 2*, by interaction with the natural environment, outputs are produced to meet the needs of all of the members of the community. Part of these outputs will be in the form of producer goods to facilitate further economic activity. Such assets enable increased levels of output. Stewardship comes into this picture at each of the key points: work, resources and needs. In relating *work and resources*, stewardship involves not only effort but also judgement. *Effort* and ability, in the sense of technical *efficiency*, can be wasted if misapplied, and ability in the sense of *judgement* is fundamental to ensure that outputs of a suitable *kind* are produced. Performance measurement, allowing for judgement, would become very subjective and distributive justice on this basis would eventually become a matter of individual calculation. Reliance upon the market to reward individuals removes the subjective element and the scope for injustice arising from human error of assumption or calculation. It provides rewards related to consumer perceptions of their needs and the extent to which respective producers efficiently meet these needs. Reliance upon the market in this way does not have to become an ideology. It is no more than the acceptance of a mechanism that most of the time can deliver approximate distributive justice when alternative calculations are likely to be time consuming, resource consuming, subjective and at the end of it all, inferior.

Social Utility

In connection with *resources and need*, we have seen in *Chapter 3* and *Chapter 4* that the stewardship concept is of use in addressing need especially family welfare. We have recognized that the application of stewardship to the latter cannot be confined to material goods. Debt difficulties draw attention to the fact that total family stewardship embraces such things as health and relationships. We saw in *Chapter 3* that responsible stewardship prioritizes long term domestic stability over short-term objectives and the investment objective over immediate consumption. This has a bearing on *social utility*. The family remains the core unit of society and, as such, an important focus in connection with what is useful to society.[8] When debt crisis breaks up the family, the costs of child care pass to the community, the resources consumed on housing may well be doubled and the bad debt cost ends up being borne by other groups in society. It is, therefore, apparent that those things that contribute to long term family welfare are in the interests of the society. It would be intrusive to allocate resources to families according to what others considered appropriate. Adults expect to make their own decisions about what is best for them. Some save and invest in owner-occupation. Others spend.

Inequalities of asset distribution thus arise from alternative strategies, rather than inadequate access such as arises when people in poor areas are unable to obtain mortgages. Taking the total picture we would conclude that better domestic stewardship would tend to greater social utility by minimizing the social costs resulting from family breakdowns. But effective domestic stewardship requires that the family members know what they need, as opposed to what others would persuade them to buy, and that there is access to the relevant goods. It presupposes that purchases result in the right to retain property, and that effort and merit have claims to inclusion as well as equality. It does not need to look to the market as a complete distributive system, but recognizes its capability to facilitate mutually acceptable exchanges of property. With these criteria in mind we apply the stewardship concept to mortgage finance.

5.2. Mortgage Finance

5.2.1. MORTGAGE FINANCE AND CRITERIA OF JUSTICE

As we have seen "Rent/Mortgages" was one of the most important items of expenditure in connection with the debtors in the *Jubilee* survey. In absolute terms, mortgage finance dwarfs the other categories of borrowing. Analysis of total personal sector credit in 1989 showed that mortgage finance represented 65% of borrowing with bank lending accounting for only 8% and trade and other credit 23%.[9] The importance of mortgage finance is further emphasized by the fact that mortgages outstanding have increased over four times in less than a decade, rising from £62 billion in 1981 to £256 billion in 1989.[10] These funds are borrowed in connection with an investment objective and the outcomes for the majority of borrowers are favourable. The figures show that house purchase through mortgages involves a massive distributive exercise through the property and financial markets, which results in legal entitlements to houses shared between mortgagors and mortgagees according to the loan terms.

This suggests a linkage between rights and responsibilities. If there are to be *rights* to own, these must be balanced by responsibility. As we have seen above, responsible stewardship implies *effort* and *ability*.[11] Both are relevant to mortgage finance and it might be argued that all should not expect an *equality* of access to ownership through mortgage funds if they are not prepared to accept the same responsibilities of personal sacrifice to make repayment.

The *need* for accommodation can be addressed in different ways. When individuals or families take advantage of mortgages to finance home ownership through the housing market it is of considerable *social utility*. Public funds are not required in order to meet the accommodation needs represented. However, mortgage default has risen sharply through the 1980s. According to the building societies their property repossessions increased almost fivefold in six years from 4240 in 1981 to 20550 in 1986.[12] By the second half of 1990 more than 1000 homes a week were being repossessed and at the end of 1997 it was 800 per week, with only 1 in 4 taking mortgage protection. The Council of Mortgage Lenders reported repossessions in 1990 as 44,000.[13] These repossessions bring personal distress and social costs and it is essential that home ownership through mortgage finance is linked to the kind of domestic stewardship that we have argued for in *Chapter 3* if distributive justice is not to be thrown into chaos by extensive debt default.

5.2.2. MORTGAGE DEBT MANAGEMENT

Stewardship and Owner Occupation

One measure of *development* is asset enhancement and mortgages provide an opportunity for such enhancement if assets appreciate. Borrower solvency can improve, even if the mortgage is allowed to increase, if assets appreciate at high rates. High-multiple, low start schemes provide an example of how borrowers can cash in on this. A solicitor or architect might borrow as much as four times normal salary. Interest might be 2% above the standard mortgage rate but, with interest deferral on a sliding scale for the first five years, the repayments would be less than for someone borrowing up to 3.25 times their salary with no low start. The increase in the mortgage in consequence of deferred interest could be compensated for by price inflation, allowing owner equity to increase, by as much as the amount of the loan, in only 5 years.[14] It is not surprising, therefore, that property acquisition should be viewed as a primary objective.[15] In general our stewardship concept would support this, but not unreservedly. As we have seen in *Chapter 3*, the concept supports the pursuit of goals that improve solvency but this is not the same as saying that moral decision can be reduced to a simplistic assumption, as there is the problem of the complexity of the market. Even with careful stewardship the market may deliver unexpected results, so there are a variety of additional considerations to be addressed in the following sub-sections.

Asset Value Enhancement and Market Morality

Consideration of asset value *enhancement* in the case of property might suggest that the market rewards good stewardship and that it possesses an inherent justice. There have been attempts to interpret the capitalistic work ethic in such a way in defence of contemporary market theory. Irving Kristol, for example, presents a defence of capitalism on the basis that equality of opportunity will result in everyone's betterment, although not to an equal extent. He justifies inequalities on the grounds of unequal contributions.

> "... some people end up with a lot more than they had and some with a little more than they had. Capitalism does not perceive this as a problem. It is assumed that since everyone gets more everyone ought to be content. If some people get more than others, the reason is to be found in their differential contributions to the economy."[16]

On this understanding of the market economy, the better our stewardship, the more we will prosper. The system:

> "rewards people in terms of their contribution to the economy as measured and defined by the market-place - namely, in terms of the free preferences of individual men and women who have money in their pockets and are free to spend it or not on this, that, or the other as they please. Economic justice under capitalism means the differential reward to individuals is based on their productive input *to the economy*."[17]

To endow the marketplace with such a capacity for economic justice is tempting, because, in general terms, there is a correspondence between effort, diligence, correct judgement and returns. But as the housing market shows, market outcomes can be different to what we expect. Mortgage purchases of domestic property, which had all of the appearance of shrewd investment turned sour on many borrowers.

Stewardship Goals and Distributive Justice

Are the disappointments to be accounted for on the basis that home ownership is not an appropriate consumer goal? Or is it just that it was not appropriate in the particular cases where things went wrong? In developing our personal stewardship model in *Chapter 3* we emphasized the importance of the consumer comparing goals with the position disclosed by a self-assessment, with a view to matching expectations to financial, physical and human resources. Our stewardship approach would certainly recognize that in answering the question whether or not home ownership is an appropriate goal, it is necessary to look at the circumstances of the particular borrower. But the latter might regard such a goal as belonging *equally* to everyone and seek support from the general conception of justice referred to by Rawls.

> "All social values - liberty and opportunity, income and wealth, and the bases of self-respect - are to be distributed equally unless an unequal distribution of any, or all, of these values is to everyone's advantage." [18]

Not all philosophers would agree with such a conception. J. J. C. Smart, for example, accepts that, in practice, utilitarianism may tend to egalitarianism but states clearly that its main concern is with maximizing total *happiness*.

> "A utilitarian will hold that a redistribution of the means to happiness is right if it maximizes the general happiness, even though some persons, even the least advantaged ones, are made worse off." [19]

This distinction of viewpoint is significant for lending practice. These two views raise the issue as to whether all should be on an equal footing or whether it is adequate that *most* benefit even though *some* are made worse off. We have already seen that this distinction is relevant to credit scoring. A cost-effective system will serve the bank and *most* customers adequately in preventing default but it will be at the cost of *some* casualties, some of which could be avoided by employing staff intensive procedures for loan approval. Our stewardship approach would recommend seeking to prevent the casualties and the same would be true in relation to mortgage finance.

Supplier Pressure on Borrower Decisions

In *Chapter 3* we saw the importance of *matching* with respect to expectations and goals. External pressures can have a significant effect upon the realism of such stewardship measurement. We have recognized above that there is no question that *need* is involved in seeking accommodation. The problem that arises is that what is thought necessary may be the result of external influences that mislead the individual with respect to what the true necessities are. John Kenneth Galbraith has addressed this issue from an ethical perspective. He has argued that the process of production in fact creates many of the unsatisfied wants in modern society. In explaining this "dependence effect", Galbraith argues that wants cannot be urgent if they are contrived.

> "... above all they must not be contrived by the process of production by which they are satisfied. ... One cannot defend production as satisfying wants if that production creates the wants."[20]

There is no doubt in Galbraith's mind that the institutions of advertising and salesmanship shape the goals of the consumer by bringing into being wants that did not previously exist. He argues that in the increasingly affluent society,

> " ... wants are increasingly created by the process by which they are satisfied. This may operate passively. Increases in consumption, the counterpart of increases in production, act by suggestion or emulation to create wants. Or producers may proceed actively to create wants through advertising and salesmanship. Wants thus come to depend on output."[21]

It is interesting that increasing living standards have not led to a reduction in consumer credit; and there may be some evidence here to bear out

Galbraith's contention of a manipulation of consumer demand. As Janet Ford comments,

> "In the eighteenth and nineteenth centuries most credit was taken to manage poverty. This use of credit has remained an important one, but there has also been the development of credit related to the growth of affluence."[22]

There is research evidence to show that whereas manual workers are above average users of hire purchase and voucher trading, professional, managerial and other non-manual workers are above average users of credit sales, credit cards and bank loans.[23] Such survey findings confirm an above average experience of certain credit sources by socio-economic groups that would also be experiencing affluence. This bears out Ford's opinion that by the late twentieth century affluence-related credit developed to finance higher levels of want satisfaction. This suggests that *need* was replaced by less basic criteria. This has been particularly evident in relation to housing where families have gone far beyond what satisfied their needs in relation to living accommodation, preferring bigger and better houses. This is not, however, solely the result of demand creation.

Borrower Effectiveness Culturally Influenced

Galbraith's view is not without its critics. F.A. von Hayek has taken issue with Galbraith's dependence effect. He accepts that,

> "we would not desire any of the amenities of civilisation - or even of the most primitive culture - if we did not live in a society in which others provide them."[24]

But Hayek maintains that desire for various objects is an integral part of civilized life. There are indeed, at one extreme, basic necessities without which life is impossible; and, at the other extreme, items of conspicuous consumption, but, in between these extremes are "the overwhelming majority of goods on which civilized life rests."[25] Concerning *need* he goes on to say,

> "Most needs which make us act are needs for things which only civilisation teaches us to exist at all, and these things are wanted by us because they produce feelings or emotions which we would not know if it were not for our cultural inheritance."[26]

By way of illustration Hayek refers to works of art which would have to be regarded as of little value on Galbraith's position if

> "the fact that people would not feel the need for something if it were not produced did prove that such products are of small value..."[27]

Additionally Hayek draws attention to the fact that it is by no means proved that particular producers can determine the wants of particular consumers because their

> "uncoordinated efforts ... merely create one element of the environment

> by which the wants of the consumers are shaped."[28]

In addressing the subject of home ownership no one would want to dispute the importance of shelter. This is not something contrived but a basic necessity. What might not be original with the individual, however, is whether one should aim at renting or owner occupation and whether in connection with the latter one should resort to borrowing. Ford suggests that external non-financial factors form part of the broader environment of home ownership, particularly,

> "the beliefs and attitudes that have been conveyed to home owners by lenders and governments about the acceptability, wisdom, safety, and likely advantages of taking a mortgage and buying a house."[29]

As this quotation from Ford shows, it is possible for the judgement of individuals to be clouded by political and commercial rhetoric creating the impression that home ownership is so reasonable and desirable that to be left out is to fall short of some entitlement. Home ownership becomes almost a matter of natural human *rights* in the minds of the general public. This impacts upon credit commitment, as *access* to funds is necessary to possession.

Credit Procedures, Rights and Effective Repayment

Against this background the credit promotion policies of financial institutions assume a formative influence in connection with people's objectives. The availability of secured loans places within the reach of thousands the possibility of home ownership which would otherwise not be an option. The influences are complex. 100% mortgages remove the constraint of finding 10% deposit enabling borrowers to set their sights on higher value properties. Basing repayment ability on joint incomes and increasing the salary multiple used can have similar effects on consumer goals. Tom Sorell and John Hendry draw attention to the fact that the NCC report *Credit and Debt* contains evidence that:

> "the growth in indebtedness in the 1980s was associated with the growth in the supply of credit for housing combined with a lowering of the requirements for lending from established sources of credit, such as building societies."

With respect to the latter they cite:

(a) the growth of the number of loans covering all or more than the purchase price,

(b) the growth in loan amounts calculated using annual income multiples, and

(c) the decline in the quality of the mortgaged property.[30]

Rule of thumb criteria can convey the impression to borrowers that is safe to take the *maximum* advance available; and they may adjust their property acquisition objective accordingly. Entitlement features in credit promotional materials, leaving the impression that at low levels of commitment borrowers are taking less than their *right*.

Loan Purpose: Speculation or Stewardship?

Mortgage finance can result in speculative ventures rather than good stewardship. Instead of *need*, *effort* and *social utility* having a prominent place in the decision taking, a situation can be created where reliance upon debt is seen as the route to wealth, with people extending themselves to the limit to maximize the opportunity. Tim Congdon comments on how US citizens were prepared to go heavily into debt to make speculative fortunes.

> "They borrowed from their banks at interest rates of under 10 per cent in order to purchase assets appreciating at 15 per cent a year. The past record of continuous asset price appreciation encouraged expectations of further appreciation in future..."[31]

He explains how by the early 1970s there was a growing demand for mortgage debt that the financial institutions were keen to satisfy. The consequent expansion in mortgage borrowing fuelled an even higher rate of house price increases encouraging optimism about continuing price increases. He quotes by way of example the five years to 1978 during which the price of new one-family homes rose by 11.4 per cent per annum. During the same period the pre-tax cost of mortgage finance was in single figures.[32] He maintains that by the late 1970s:

> "most Americans took it for granted that borrowing to buy a house would be profitable, in the sense that the increase in its price would exceed mortgage costs ..."[33]

There is an issue of economic justice here. Tax deductibility of interest paid on mortgages benefits borrowers whereas those not borrowing lose out. There is a measure of income redistribution from the latter to the former. It is therefore no matter of surprise that the tax considerations stimulated the appetite for mortgages throughout the 1970s and early 1980s. A person with a marginal tax rate above 50%, taking on a 10% mortgage, would pay a post-tax interest rate of 5% to acquire assets rising in value at 15% per annum. As inflation was leading to increased incomes moving more personal taxpayers into higher tax brackets, more taxpayers could benefit from the tax relief and the demand for credit strengthened. Congdon goes on to explain how,

> "A tax avoidance industry ... successfully preached the beneficence of credit to an American middle class which had believed for three centuries in the Puritan virtues of hard work and thrift."[34]

He elsewhere comments,

> "Whereas in the 1940s common prudence and recent memories obliged people to accumulate financial assets in order to avoid bankruptcy, in the

> late 1970s the fashion was to borrow heavily, acquire rapidly appreciating assets and profit from inflation."[35]

This USA situation illustrates that credit availability and economic conditions influence both consumer objectives and consumer strategies. The impact of mortgage relief is significant in this connection. Not only does it provide an opportunity for house purchase far beyond *need*, the subsidy becomes a lure into levels of commitment beyond the borrower's capacity to repay. Were justification for it to be sought in relation to distributive justice, we might expect reference to be made to *social utility* or *need* by way of justification but the scenario Congdon describes produced a situation where debt dependency was preferred to saving, not on the basis of these criteria but as the route to riches.

5.2.3. DEBT DEPENDENCY

Asset Depletion: From Savings to Equity Withdrawal

We noted in *Chapter 3* the centrality of a *stewardship strategy* for total family welfare and saw that saving is an alternative to reliance upon debt. There is evidence to suggest that in the UK the 1980s witnessed a shift from the savings strategy to the borrowing strategy with its dependency upon future income instead of past savings. The second half of the 1980s saw a fall from 20% to 15% in total private savings as a percentage of GDP at factor cost.[36] Total national savings remained stable only because public sector saving compensated for this reduction. As a percentage of personal income the fall in savings during the decade was significant. In 1980 it stood at 15 per cent, by 1985 it had fallen to 9.5 per cent and in the late eighties it had reached one third of the 1980 figure at 4 to 5 per cent.[37]

Alan Davies mentions two relevant factors. The first has to do with attitudes to the future and the second with mortgaging that future. With respect to the former he says,

> "...to the extent that households believed that their future income would be higher than they had previously expected ... then the effect would have been to raise consumers' expenditure and lower the saving ratio. Borrowing made possible by financial deregulation may be seen as a channel in this process rather than a causal factor."[38]

If this is a correct analysis it would seem that consumers were not concerned to employ the savings strategy to provide for the future. Rather, they were happy to leave the future to look after itself, or, by utilizing the borrowing strategy, they sought to cash in on the future to secure present enjoyment or asset acquisition. This brings us to the second factor, of which Davies says,

> "Soaring asset values (notably house prices) together with, from 1987, the

ability of home-owners to raise loans for most purposes over an extended term against the security of their property, encouraged individuals to consume part of their increased wealth via so-called equity withdrawal."[39]

These comments illustrate the point made earlier that there is an interaction between the lending policies of financial institutions and consumer objectives. Borrowing facilities to enable equity withdrawal provided homeowners with further opportunities to secure present consumption on the strength of asset values and lender claims against their future earnings. Congdon reports a similar phenomenon in the USA in the 70s. As borrowers' equity values increased,

"... they could offer better security for other kinds of consumer credit. The house price boom of the mid-1970s was therefore accompanied by buoyant credit for the purchase of cars and consumer durables."[40]

Against such a background it might seem unlikely that spenders will pay much heed to the comment of Neil Collins, City Editor of *The Daily Telegraph* to the effect that:

"it no longer makes sense to borrow to buy something, and let inflation take care of the debt. The incentives offered to savers are such that 'saving up to buy', a practice destroyed when prices rose faster than savings ever could, makes sense again."[41]

The problem is that exhortation can be ineffective in the face of the economic climate created by government intervention and institutional credit policies. J.G. Dees has issued a similar caution with respect to charity. He emphasizes the importance of looking at the *specific* economic impacts in the *particular* cases. He explains as follows.

"...charity in any form may seem like a good thing. It pleases the giver and benefits the recipient. However, economic analysis may raise serious questions about the consequences of certain forms of charity. It may harm the recipients by creating incentives for dependence rather than self-sufficiency."[42]

The relevance with respect to family stewardship and strategy is addressed in the following section.

Government Policy and Economic Justice

Successive Chancellors have sought to encourage the savings strategy in a general way. Save as you earn schemes have been in operation since 1969. In 1980 a share option related save as you earn scheme was introduced. Personal Equity Plans were introduced in 1986 and personal pensions in 1988. In 1990 the Chancellor announced Tax Exempt Special Savings Accounts and with the abolition of the composite rate in 1991 non-taxpayers could receive interest gross on their bank accounts. David F. Lomax sees such schemes as a necessary contribution to the individual being able to protect themselves against times of adversity in an increasingly risky environment with redundancy and employment re-training becoming more common.[43] The problem remains, however, that the present system of tax relief on mortgage interest encourages borrowing, while government interest rate and fiscal policy discourages saving for taxpayers. The abolition of composite rate tax is instructive. While 15 million savers on low incomes stood to gain by the abolition of the 22 per cent composite tax, 26 million income tax payers stood to lose by having an additional 3 per cent tax deducted.[44] The savings and loans market illustrates that, at the moral level, government policy initiatives, which have the appearance of distributive justice, involve, often in a large number of *particular* cases, economic *injustice*. Our stewardship model at least has the advantage of providing a basis to address each case individually avoiding the kind of economic injustice that we have been discussing.

5.3. Money Advice

5.3.1. ADVICE SERVICES AND ACCESS

Issues of justice and procedural fairness arise in money advice of which the heart is supportiveness. Default illustrates the need for help with financial management. We have already seen that *access* is important with regard to distributive justice. There has been a growing recognition of the problem of access to effective money advice. Enquiries to London *Citizens Advice Bureaux* about consumer and debt issues rose from 110,000 in 1979/80 to 156,000 in 1987/88 [45] and a variety of alternative providers are now well established.[46] About the same time *Citicorp* undertook to fund the initial of operation of a London-wide *Citizens Advice Bureaux* debt advice unit to support the 1300 advisers.[47]

The *National Consumer Council* and the *General Consumer Council for Northern Ireland* were involved in Lord Ezra's committee to consider money advice services in the UK [48] and the *NCC* undertook a survey of present provision. The report *Debt advice provision in the United Kingdom*, by Tim Young was published in May 1990. Lord Ezra's working party also published its report in 1990. The *Money Advice Funding Working Party* found that serious arrears affected 200,000 households and that because of piecemeal development access to adequate money advice coverage was lacking. It concluded that good money advice organization would benefit credit granters and that a full review was needed.[49] It recommended that:

"a thorough review of the money advice organisation should be

> undertaken to ascertain the most cost-effective way of providing advice to the general public."[50]

The problem of access is real as is the need for cost-effectiveness of provision and in responding to the problem it is important to have in mind the distributive justice criteria we have already elaborated in *Part I* such as social utility, need, effort and ability.

5.3.2. STEWARDSHIP VALUES IN MONEY ADVICE PROVISION

Against this background we will consider what our stewardship model would recommend in connection with money advice provision. Key ideas that we have developed in relation to entrustedness include *supportiveness*, *fairness*, *co-operation* and operational *effectiveness*. Supportiveness will be taken up first in relation to social utility.

Supportiveness and Social Utility

Complexities in lending require appropriate support for borrowers. Increased provision of financial information by lenders has not solved the lack of consumer awareness about the cost of credit including total paid in interest and related charges, flat rate interest rate and annual percentage rate.[51] In 1988 a *Public Attitudes Survey* found that only one in five consumers knew what APR signified.[52] We mentioned earlier the *social utility* arising from the reduction in default and we saw in *Chapter 4* that to avoid debt casualties *supportiveness* is necessary before borrowing takes place. The need for preventative action is recognized. The London *CAB* debt advice unit identified educational and preventative work as a necessary part of its operation. The press has conveyed this message to the general public.

> "Financial deregulation means the end of the straitjacket of mortgage quotas, hire purchase controls and under-rewarded savers. But freedom means freedom to get things wrong, which means that education, understanding and information is vital in money matters today."[53]

Progress has been made in providing information. Examples include:

(1) *Office of Fair Trading* pamphlets in a popular style to give debtors the basics of an action plan working out deficits, identifying amounts owed and communicating with creditors.[54]

(2) Local Authority materials produced by *Trading Standards Departments* and *Public Relations Departments*, such as *Cleveland County Council*'s: *Getting into credit is easy* and *Getting out of debt is hard* covering advice on APR, written quotations and types of credit available, sorting out debt problems and addresses of centres where debt advice is available.

(3) Newspapers often include special advice. The *Independent*'s Spring 1991

Student Life, for example, included an article entitled 'Tax-free way to profit from debt'. This was an examination of how to make the best use of the student loans scheme. [55]

(4) Voluntary societies such as *Credit Action* which has produced a money management course, *Mind over Money* with practical advice for those struggling with debt.[56] Sessions deal with (a) budgeting, banking, and money morality, (b) planning future finance, budget control, saving, and the ethics of investment, (c) using money, cash, credit and giving, and (d) cutting losses.[57] The course provides some basics for money management so that funds can be repaid and it does so in an ethical context.

The previous chapters of this book have shown the importance of considering technical debt management operations in a moral context and have shown that the stewardship concept is a valuable aid to such analysis. As has been seen in this section our inclusion of *supportiveness* is consistent with the criterion of social utility.

Procedural Fairness, Effort and Need

We said at the outset of this chapter that we would discuss both mortgage finance and money advice in the light of distributive justice. By considering stewardship in the context of economic activity we saw how it involves not only *access* and *social utility* but also *need*, *ability* and *effort*. These last three come particularly into focus in the debt advice situation because there is a balance that needs to be preserved between *need*, on the one hand, and *ability* and *effort* on the other. This arises from the need to be fair to both creditor and debtor. Gregory Dees identifies four areas relevant to fairness in contractual relationships:

> "the process used to reach agreement, the terms of the agreement, the way in which it is implemented, and the outcomes it creates."[58]

We will address Dees's four areas: process, terms, implementation and outcomes, using our stewardship model, while making particular reference to balancing the distributive justice criteria: *need* and *effort*.

The Process of Debt Advice. Fairness and *co-operation* are important for debtors as imbalance may arise disadvantaging debtors or creditors. Berthoud helps us to understand debtor problems. Studies of money advice clients suggest that debtors were confused by the sequence of demands from their different creditors not realizing they were involved in negotiating. He explains their reactions as follows.

> "They assumed that the courts would automatically take the creditors' point of view. So their instinct was to conceal the extent of their

> commitments. They tended to offer the whole of their available income in response to each demand. But the same money might be promised to another creditor the next week ... Obviously it was not possible to keep these mutually incompatible promises. Creditors understandably got the impression that the debtors were dishonest, and ready to evade their obligations."[59]

These findings illustrate the way in which perceptions about fairness, which may not be correct, result in deteriorating relationships. Debtors must be assured that debt will not place them beyond the pale of fair treatment if only they communicate honestly with their creditors about their *need.* Creditors ought to be mindful that cynicism about the debtor's intention to repay may be built upon nothing more that the symptoms of debtor inadequacy in point of *ability* to handle the situation.

Money advice providers have been mainly local authorities and the voluntary sector not the institutions as the *NCC* report *Debt advice provision in the United Kingdom* [60] shows in its summary of the providers of debt advice services. Within the statutory sector the local authority role was found to be predominant. A small number provide money advice services directly to the public using *Money Advice Centres* located within *Consumer Services* or *Trading Standards* departments. Such centres are also involved in taking cases on referral, back-up consultancy and training. Specialist debt advice workers may also be employed in specific departments to provide advice, back-up consultancy or training. A typical example would be within the *Housing Department* dealing with rent arrears problems. Some local authorities that do not employ specialist debt advice workers employ staff whose job includes giving assistance to clients in connection with personal debt problems. The report explains,

> "This assistance may vary from identifying that a client has a debt problem and making a referral to another agency; offering preliminary information on how a client should go about dealing with a debt problem; or offering advice and assistance that ranges from basic debt identification and checking, through budgeting and income maximisation to calculating repayment schedules, negotiating with creditors and representing clients in court."[61]

Local authorities may also ensure the provision of debt advice by funding voluntary sector advice provision. The pattern of provision is diverse, depending upon locality and resources. The *Citizens Advice Bureaux* have a significant role but other agencies are also active such as the *Soldiers' Sailors' and Airmen's Families Association* (SSAFA). Money advice centres in the voluntary sector provide a service directly to the public but may take cases by referral, provide back-up consultancy and training.[62]

We may not, however, prioritize *need* to the exclusion of *effort* and *ability*. The practicalities of money advice reveal that a balance must be struck between creditor and debtor responsibilities. The borrower has been entrusted with the loan funds and therefore has an obligation to repay. The process of money advice should seek to strike a balance between the creditor's responsibility to contribute to the alleviation of the borrower's debt burden and this obligation to repay. The process of money advice must involve a reversal of the irrational panic response of the debtor that Berthoud identified, so that constructive action can take place involving appropriate *effort* on the debtor's part. Berthoud summarizes the adviser's task as follows.

1. Inform all creditors so that they stop demanding non-existent funds.
2. See if income to meet expenditure and repayments can be increased by benefits and determine how much can be available to meet debts.
3. Make arrangements with creditors who have considerable sanctions at their disposal.
4. Negotiate regular payments with the remaining creditors .[63]

It takes time but by this method an acceptable repayment schedule can be set up for the average multiple debtor [64], to improve debtor *ability* to control finances recognizing debtor *need* as well as creditor claims and aiming to enhance debtor *effort* to effect repayment.

The Terms of the Agreement: Fairness and Debt Forgiveness. As we saw from Dees, procedural fairness involves the terms of the agreement and the process outlined above raises the question of the terms of repayment and particularly, how much should be repaid. We will address this in relation to debt forgiveness. It is within the context of a fair balance between creditor and debtor that the subject of debt forgiveness arises. It might be argued that the lender ought to give interest relief. Indeed, when borrowers get into difficulties, is not the debtor *need* so great that the lender ought to forgive the debt? Often, such generalized statements appear fair but they obscure the complexity of decisions concerning justice in particular cases. Solomon points out how such decisions involve contribution, ability, needs, responsibilities and so on.[65] No generalized formula or theory can work without consideration of the particulars of the individual case being addressed. A broad-brush approval must be subjected to more detailed analysis in the light of stewardship entrustedness. The basic concept involves accountable control of resources. The borrower has been entrusted with the loaned funds. The lender has made available the funds and has given the borrower control for a specified purpose. The borrower's objective might be to purchase or renovate a house or to invest in a business. The funds are not given as an outright grant and the borrower is accountable to to repay the funds and interest charges. The lender expects that the borrower will manage his financial affairs in an efficient manner so that repayments can be made on the due dates. Interest relief may arise when its payment would make loan reduction impossible. Debt forgiveness of small amounts save collection and court costs making it attractive to lenders. However widespread expected debt forgiveness would undermine the whole financial system if both lenders and borrowers fail to take seriously the demands of stewardship.

What guidance do the social responsibility implications of the stewardship concept provide? We have seen that stewardship control of resources necessitates effectiveness. When measured financially, effectiveness is indicated by good profits. Writing off bad debts reduces profits so managers should only write off debts to save further losses. However, our expectation that lenders will recognize that their accountability to borrowers requires them to justify their actions in the light of all the circumstances involves more than profit maintenance. If, for example, investigation into the default reveals bank actions contrary to the principle of supportiveness, then this should be allowed for. A borrower might be given inappropriate advice at the time of taking the loan. If such matters were found to be relevant then the principle of fairness would dictate that profit should not be given absolute priority. The principle of contribution could be usefully applied and the extent to which bank staff had contributed to the debtor's default could be used in determining the level of debt forgiveness applicable.

Forgiveness, however, implies that the debtor is responsible for the default. Forgiveness is gratuitous release of one in a position of demerit. There are doubtless cases where it might be determined that the bank was not at fault and that the debtor experienced no particular adverse circumstances. Stewardship matching might show, for example, that the debtor had sufficient income for repayment but failed to exercise ongoing control and engaged in impulse buying of luxuries. Such a position would require great caution even if the borrower were hopelessly in debt. In connection with attempts to apply a *Jubilee*-style cancellation of debt, Peter Selby explains the repeated emphasis upon securing a fresh start for deserving debtors without undermining the general obligation of repayment. Speaking of sovereign debt he says:

> "As a result of this view of the sanctity of debt it appears that the most likely way forward in practice will be a case by case consideration of the needs of particular countries. In the manner of applications for bankruptcy, where there is a genuine case that the burden of debt is clearly beyond repayment, an opportunity for a fresh start will be given and full payment will not be required."[66]

The stewardship concept does not limit the issue to maintaining debtor obligation to repay. It also takes up the future orientation. If a fresh start is given, it would want to question the kind of new start that results. In the case of default caused by debtor mismanagement, a new start without new principles would lead to a repetition of the old problem. Certainly it would be irresponsible on the part of bankers to disregard the causes of the original default and, writing off outstanding debts, restart on the same inadequate basis. This would be detrimental to the maintenance of stewardship entrustedness on the part of the bank.

The argument that because banks are large and impersonal the responsibility of repaying is less than where another individual is the lender is weak. The reluctance of banks to engage in extensive write offs is not only a question of profits. As Morris Miller explains:

> "For the commercial bank creditors the most important issue has been the impact of such defaults or forgiveness on their own solvency and bottom line, and for their Governments, the risks for the global financial system."[67]

O.A.W. Ragkhan came to a similar conclusion in his thesis *The international debt crisis and the economics of Islam.* While approving of debt forgiveness on the basis of the Qur'an (*2:280*) he recognized the practical difficulties posed by the threat to bank solvency.[68] It is naive to think that bank losses or casualties only impact on the rich. It is the stakeholders who bear the cost. Unpaid debts lead to reduced returns to shareholders, depositors and staff and higher costs for customer lending. In contrast effective operational stewardship benefits the whole community resulting in a just outcome. Not only is careless writing off contrary to the banker's trusteeship, it would encourage laxity among borrowers in connection with stewardship effectiveness. If expectations are raised that the bank's shareholders, depositors and staff will bail out defaulting borrowers, despite their own ineffective financial stewardship, then the necessary seriousness concerning financial planning and control is likely to diminish with a declining sense of urgency about the economic realities of repayment. The application of the stewardship concept encourages balanced ethical decision taking in debt management, securing justice for all of the parties involved. There can be no short cuts. In each particular case it is necessary to strike a balance between creditor *contribution* and debtor *need* and also between debtor *need* and debtor *effort* if we are to be left with a financial system that has any long- term integrity.

Implementation: Effectiveness in Debt Advice Provision. Dees identified fairness in implementation as the third area for consideration and this is important in money advice. It is important that the creditor receives an appropriate level of repayment in the light of all the circumstances and that the debtor has a repayment plan that is realistic and lifts the family out of crisis. It is also important that the process by which this is secured is not unduly expensive for whoever funds it, whether private or public sources. In this section we will consider the issue of *effectiveness* which we have seen, in previous chapters, is raised by the stewardship debt management model. Money advice, like any other need, can be under-resourced or over-resourced relative to other needs. We defined entrustedness as accountable control of resources and it is clear that money advice provision, like the provision of other goods and services, will be wasteful if not effectively managed. It is important, therefore, that real benefits arise from the activities. Russell Mannion and Gillian Parker of the *Social Policy Research Unit* at *York University* have researched the benefits of money advice services. The research set out to evaluate (a) the net financial costs and benefits to creditors, (b) the financial, social and psychological benefits to clients, and (c) the efficiency and effectiveness of the support unit model for developing money advice.[69] The findings as published in November 1990 were as follows.

> *Rent arrears* of tenants with advice increased by only 25%. (Others 100%)
> *Debt counselling* helps avoid eviction or repossession.
> *Specification of the client's financial position* by advisers puts creditors in a better position to pursue the most cost-effective means of debt recovery.
> *Debt repayments* by money advice clients are not less than those of other defaulters and do not create additional administrative costs for creditors.
> *Client financial, social and psychological well being* improves.
> *Repayment plans* for unemployed people can result in a 'poverty trap'.
> *Money advice support units* providing back-up services for complex cases, courts, creditor liaison and media promotion is a useful model.[70]

This research showed that money advice could be beneficial to debtors, creditors and third parties such as local authorities. A major problem is that there are different kinds of debtor and the same support provision is not necessarily appropriate to every case. The stewardship model that we have advocated is clearly relevant where the kinds of recommendations that were suggested in *Chapter 3* can be implemented. But in some cases the problem is not effective debt management but poverty. At some income levels, no amount of improvement in debt management is going to solve the problem. *Need* has gone beyond what *ability* and *effort* can restore. Matching of income and outgoings merely reveals the irreversibility of the deficit and the impossibility of repayment. The need is not for *advice* but *finance* and mechanisms other than debt management are required.

In a different context, Thomas Donaldson, addresses the ethical problem involved, that, when stipulating what a debtor ought to do, the debtor's *ability* is relevant. In developing his theory of rights Donaldson employs a "fairness-affordability" test. By this he means that:

> "all moral agents ... must be able under ordinary circumstances, and after receiving any share of charitable help due them, to assume the various burdens and duties that fairly fall upon them..."[71]

Applied to default situations this would mean that where there is no capacity to repay at present or in the future, laying an obligation to repay upon a moral agent would be self-defeating. Either the debt must be cancelled or additional income must be found. This has been recognized in the operation of the highly successful *Consumer Credit Counselling Service.* The service makes a distinction between those debtors whose problem is poverty, and who require to be referred to agencies capable of addressing this need; and those debtors whose problem is debt management, who, as clients of *CCCS,* will benefit from one of their *Debt Management Plans*. For multiple debts, the plan provides for one monthly payment to *CCCS,* which is allocated to creditors on the basis of the relevant contracted monthly payments for the debts. The underlying ethic supports at several points the application of stewardship values. It implies *entrustedness* in upholding repayment where there is an ability to repay through rescheduling of debts. It recognizes the need for *supportiveness* in the light of widespread weakness in financial skills among defaulting debtors. It focuses upon *matching income and expenditure* by relating monthly household income to essential living expenses and priority debt payments to arrive at the monthly payment. The *Debt Management Plan* is designed to secure *effectiveness* in repaying creditors.[72] The *CCCS* model has been popular with both debtor clients and participating creditors, and its success suggests that stewardship values would be acceptable to borrowers, lenders and advisers in the implementation of debtor rescue packages.

Outcomes: Accountability to stakeholders. In terms of our stewardship model outcomes are addressed through accountability. Accountability is relevant in relationships between debtors and creditors and debt advice provision. Trustee management involves accountability to the various stakeholders and it can be argued that, if the banks are expected to make contributions to finance money advice services, they are entitled to expect that the funding will be efficiently used. It is clear from independent findings that there is scope for improvement in the service. The *National Consumer Council* report on debt advice provision included comments on how the provision of debt advice might be improved in the future.

1. A national strategy is needed to address widespread personal debt.
2. Government, local authorities and creditors should provide adequate, secure and long term funding.
3. Organisation should include an industry/ advice forum to discuss policy, a national network of front-line agencies, specialist back-up

facilities and adequate liaison between all agencies and courts.

4. A nationally accredited training scheme for all money advisers.
5. Better publicity about the available advice agencies.
6. More attention to prevention. Lenders should be more careful in ensuring that borrowers will be able to pay and impress upon them the need for insurance cover. More education on money and debt.[73]

The complexity of any resulting future organisation will be clear from these observations and the complicated nature of such arrangements would make accountability concerning cost-effectiveness very problematic. The *CCCS* model appeals to creditors because the voluntary contribution of 15% of the amount repaid involves payment by results. This preference suggests support for *accountability* and indicates that the repayment *outcome* as a measure of proportionate funding is acceptable to most lenders. We have stressed the importance of measurability regarding effectiveness, and the *CCCS* model illustrates how the repayment outcome can find acceptance where considerable expenditure on bureaucratic and subjective apportionments would not. Effectiveness is demonstrated by the repayments and disputes about relating funding to benefits are superfluous If the *process*, *terms* and *implementation* are fair, as explained above, we can see how, even in complex situations, the stewardship principle of accountability can be related to *outcomes* in a practical and successful way.

5.3.3. CONCLUSION

Our stewardship approach would provide a suitable basis for a national strategy supported by lenders, consumer organizations and advice agencies. We have related our ethical discussion to their views. Present initiatives emphasize preventative education and debtor support and the stewardship model is beneficial to both. Borrowers using our model are less likely to default reducing the burden on advice services. The operational usefulness of the stewardship concept has received empirical support from the success of the *CCCS.* Overall we may conclude that the testing of these last three chapters shows that stewardship values provide an understandable moral core, capable of acceptance by borrowers, lenders and advice agencies, which sheds light upon operational decisions and can contribute to ethically sensitive debt management practice.

Notes to Chapter 5

1. Tom L Beauchamp and Norman E. Bowie, (eds.) *Ethical Theory and Business, Fourth Edition,* p.596.
2. E. Clinton Gardner, *Justice and Christian Ethics*, pp.133-134. See also Michael Walzer, *Spheres of Justice: A Defense of Pluralism and Equality*, pp.3-30.
3. Michael Walzer, *Ibid.,* p.227-228.
4. Robert Nozick, 'The Entitlement Theory' in Tom L Beauchamp and Norman E. Bowie (eds.) *Ethical Theory and Business, 4th Edition,* pp.612-613.
5. Tom L Beauchamp and Norman E. Bowie, *op. cit.,* p.602-603.
6. For a discussion of the philosophy of Thatcherism see Paul Heelas, 'Reforming the Self: Enterprise and the Characters of Thatcherism' in Russell Keat and Nicholas Abercrombie (eds.) *Enterprise Culture*, pp.72-90.
7. Michael Walzer, *op. cit.,* p.3.
8. The Conservatives linked self-interest and the family. See Raman Selden, 'The Rhetoric of Enterprise' in Russell Keat and Nicholas Abercrombie (eds.) *op. cit.*, p.62.
9. See Andrew Howe Browne, 'The Banks and Personal Credit', pp.17-19, *Banking World*, January,1991,
10. *Ibid.*
11. Walzer discusses rights and *abilities*. See Michael Walzer, *op. cit.,* p.152.
12. Janet Ford, *The Indebted Society: Credit and Default in the 1980s*, p.46.
13. Ian Cowie, '1,000 Homes a Week are Repossessed as Mortgage Debts Rise', p.7, *The Daily Telegraph*, February 15, 1991, © Telegraph Group Limited, London 1991. See further Duncan Maclennan, *Findings: The Future Sustainability of Home-ownership.*
14. Barbara Ellis, 'Multiple choice question', p31, *The Daily Telegraph*, September 14, 1988, © Telegraph Group Limited, London 1988.
15. See Paul Heelas, 'Reforming the Self: Enterprise and the Characters of Thatcherism' in Russell Keat and Nicholas Abercrombie (eds.) *op. cit.,* p.81.
16. Irving Kristol, 'A Capitalist Conception of Justice' in W. Michael Hoffman and Jennifer Mills Moore, (eds.), *Business Ethics: Readings and Cases in Corporate Morality*, p.55.
17. *Ibid.*
18. John Rawls, 'Justice as Fairness' in W. Michael Hoffman and Jennifer Mills Moore, *op. cit.*, p.38.
19. J.J.C. Smart, 'Distributive Justice and Utilitarianism' in W. Michael Hoffman and Jennifer Mills Moore, *op. cit.*, p.51. See further J.J.C. Smart and Bernard Williams, *Utilitarianism for and Against,* pp.3-74.
20. John K. Galbraith, 'The Dependence Effect' in Tom L. Beauchamp and Norman E. Bowie, (eds.) *Ethical Theory and Business*, pp.496-497.
21. *Ibid.*, pp.498.
22. Janet Ford, *op.cit.*, p.33.
23. *Ibid.*

24. F.A. Hayek, 'The Non Sequiter of the "Dependence Effect"', *Southern Economic Journal*, April, 1961 reprinted in Tom L. Beauchamp and Norman E. Bowie, eds., *op.cit.*, pp.509.
25. *Ibid.*
26. *Ibid.*
27. *Ibid.*
28. *Ibid.*, p.510.
29. Janet Ford, *op.cit.*, pp.74.
30. Tom Sorell and John Hendry, *Business Ethics*, p.69.
31. Tim Congdon, *The Debt Threat: The Dangers of High Real Interest Rates for the World Economy*, p.90.
32. *Ibid.*, p.173.
33. *Ibid.*, p.90.
34. *Ibid.*
35. *Ibid.*, pp.174-175.
36. Alan Davies, 'Too Much Income: too Little Saving', p.42, *Banking World*, April 1990.
37. *Ibid.*
38. *Ibid.*
39. *Ibid.*
40. Tim Congdon, *op. cit.*, p.174.
41. Neil Collins, 'Decade of the Money Revolution', p.2, *The Daily Telegraph, Personal Finance Show*, March 1991, © Telegraph Group Limited, London 1991.
42. J. Gregory Dees, 'Principals, Agents, and Ethics' in Norman E. Bowie and R. Edward Freeman (eds.) *Ethics and Agency Theory: An Introduction*, p.53.
43. David F. Lomax, 'Policy for Saving in the 1990s', p.16-17, *National Westminster Bank Quarterly Review*, August 1990.
44. See Ian Cowie, 'Less Interest for Millions', p.8, *The Daily Telegraph*, March 20, 1991.
45. 'Money Advice Initiative', p.11, *Credit Management*, May 1989.
46. See for example the *Consumer Credit Counselling Service.*
47. 'Money Advice Initiative', *ibid.*
48. National Consumer Council, *NCC Annual Report*, 1990, p.10.
49. 'Lord Ezra's conclusions', p.7, *CSA News*, Summer, 1990.
50. *Ibid.*
51. Geraint Howells, Iain Crow and Maureen Moroney, *Aspects of Credit and Debt*, p.48.
52. *Ibid.*, p.84.
53. See p.2, *The Daily Telegraph : Personal Finance Supplement,* March, 1991, © Telegraph Group Limited, London 1991.
54. Office of Fair Trading, *CreditWise: Your Guide to Trouble-free Credit,* and *Debt: A Survival Guide.*
55. Lorna Bourke, 'Tax-free way to profit from debt', p.4, *The Independent: Student Life*, Spring 1991.

56. Credit Action, *Mind Over Money: Participant's Notes,* p1.
57. *Ibid.*
58. J. Gregory Dees, 'Principals, Agents, and Ethics' in Norman E. Bowie and R. Edward Freeman (eds.) *op. cit.*, p.41.
59. Richard Berthoud, *Credit, Debt and Poverty*, p.35.
60. National Consumer Council, *Debt advice provision in the United Kingdom: A report by Tim Young of the Community Information Project.*
61. *Ibid.*, p.3.
62. *Ibid.*, pp.3-4.
63. Richard Berthoud, *op. cit.*, p.35.
64. 'The View from the Inside: How Money Advisers Deal with Debt", p.4, *CSA News*, Undated.
65. Robert C. Solomon *Ethics and Excellence: Cooperation and Integrity in Business,* p.239.
66. Peter Selby, *Grace and Mortgage: The Language of Faith and the Debt of the World,* p.100.
67. Morris Miller, *Debt and the Environment: Converging Crises,* p.153.
68. O.A.W. Ragkhan, *The International Debt Crisis and the Economics of Islam,* p.117.
69. 'Evaluation of Money Advice Services', p.11, *CCA News*, Jan./Feb., 1990.
70. *Social Policy Research Findings no.10*, York: Joseph Rowntree Foundation, 1990.
71. Thomas Donaldson, *The Ethics of International Business*, p.75.
72. 'Consumer Credit Counselling Service: A New Solution to Debt Problems'. Issued on behalf of the *Consumer Credit Counselling Service* by Malcolm Hurlston Corporate Consultancy Ltd. (1996). See also speeches by Malcolm Hurlston and Nigel Griffiths at FCC Concertation, 12th December, 1995
73. National Consumer Council, *Appendices to Debt Advice Provision in the United Kingdom,* Appendix 5 and Appendix 6.

CHAPTER 6

THE STEWARDSHIP DEBT MANAGEMENT MODEL

6.1. The Concept of Stewardship

This chapter addresses the need to provide decision techniques for ethical operations management. Only if a value framework can inform the decision-taking process will it secure a place in daily operations. This chapter summarizes the main findings of our research by formulating a debt management model to provide an ethical decision support tool. The first part of this book used conceptual analysis to initiate the argument. The second part applied the concept of stewardship to four different examples: small business finance, consumer finance, mortgages and money advice. In each case studied we tested our set of values in relation to debt management operations. In this chapter the argument is brought to a conclusion by constructing a specific model for debt management operations. The chapter demonstrates that the materials considered in the previous chapters and the interaction of theory and practice enable us to produce a serviceable ethical model. If it is asked whether the research is of value to stimulate ethical thinking in business operations, this work gives an affirmative answer. The answer is demonstrated by the construction of the debt management model. Discussion of the model at work will help to reveal what is of worth in it and provide opportunity to assess its importance for individuals and managers. The model is consistent with the four prerequisites found to be necessary through our analysis in *Chapter 1:* use of a reasoned approach, recognition of mutual responsibilities, reckoning with economic realities and addressing matters of trust. This consistency is possible because the model takes into consideration: financial costs and benefits, non-financial banker-customer relationship factors and community ethical factors.

6.1.1. THE FOUR PREREQUISITES

Our consideration of the debt problem in *Chapter 1* looked at the history of debt revealing the dangers of debt dependence and over-commitment. It was seen that there are various macro issues which have a bearing upon debt crises, such as interest rate levels, public debt and inflation. But it was also recognized that lax loan appraisal and advance control could be contributory factors. This raised the subject of responsible lending and the consideration of what is necessary in order to ensure ethical debt management. Four prerequisites were identified.

Employment of a Rational Approach
A model for ethical debt management must be rational, in the sense of employing a reasoned approach. This is not the same as advocating a rationalistic decision process. The latter makes the assumption that all human behaviour is rational so that it can attempt to predict outcomes. This is not the purpose of the model formulated in this chapter, which is to inform a rational approach to borrowing and lending, to counter the irrational behaviour sometimes manifested by bankers and debtors.

Recognize Mutual Responsibilities
A model for ethical debt management must recognize mutual responsibilities. Whether reference is made to "stakeholders" or "constituencies", both terms recognize parties who have interests in the operational activities of financial institutions, and no model would be complete if it provided no means of relating to the different interest groups. Our model is able to do this by emphasizing relationships.

Reckon with Economic Realities
A model for ethical debt management must reckon with economic realities or it will be a mere theoretical device of little practical use. Bankers and borrowers operate in a real world of resources, income and expenditure flows, assets and liabilities and they cannot be morally responsible if they seek to operate independently of the economic reality. Our model gives a significant place to economic matching.

Address the Matter of Trust
A model for ethical debt management must address the matter of trust in human relationships. Borrowing focuses attention upon the necessity of honouring one's commitments, being faithful to one's word and accepting accountability concerning commitments. Our model embraces both trust and accountability.

We will now look at the key values, showing how they relate to the four prerequisites and how we tested them. We will then proceed to present and discuss the stewardship debt management model.

6.1.2. STEWARDSHIP CONCEPT VALUES

The developing argument of this book has been that the analysis of the concept of stewardship provides values that shed light upon ethical issues related to debt management. It also provides components for an operational model and informs the decision process so that both borrowing and lending practice can be conducted in an ethical manner. In *Chapter 2*, the stewardship concept was analyzed to derive values related to resources and society in a language understandable to bankers, borrowers and advisers. The following values were identified: *entrustedness, matching, effectiveness, development, accountability*, and *fairness.*

Chapter 2 showed that the values can address the four prerequisites, because the concept of stewardship involves rational human action to plan the deployment of economic resources, in a way that will secure effectiveness in discharging one's responsibility to the party by whom the resources were entrusted in the first place, without damaging other interests. Our model embraces rational action, responsibility to others, resources and trust through its set of values. *When* we relate the values more specifically to the four prerequisites we see that *effectiveness, measurement* and *development* answer the need for a rational approach; *entrustedness* and *accountability* relate to responsibility; *effectiveness* and *measurement* relate to the need for economic realism; and *entrustedness*, *accountability* and *fairness* relate to trust in human relationships. This shows that the stewardship values answer to the prerequisites for an operational model for ethical debt management.

Chapters 3 to *5* carried forward our argument by testing the concept relating it to the four different examples. This provided the opportunity to relate the stewardship values to social science debt research, official reports of consumer bodies and banking literature. This achieved the following objectives:

(a) to enable an interaction between theory and practice;
(b) to relate existing social science investigations about debt to applied ethics, letting bankers and consumer bodies speak for themselves;
(c) to show the relevance and usefulness of stewardship values in thinking about debt management in an ethical and realistic way;
(d) to advance the argument by a genuinely interdisciplinary method;
(e) to present issues in a way accessible to banking practitioners and capable of engaging their interest; and

(f) to consider different *kinds* of lending from different *perspectives* such as those of borrowers, bankers and debt counsellors.

By relating stewardship values to personal borrowing objectives, the banker's canons of lending, methods of appraisal for business and consumer loans, mortgage finance, credit promotion and money advice, ethical issues have been highlighted and necessary improvements in practice identified. For instance, close analysis of individual appraisal and credit scoring, showed that stewardship values would endorse the former method rather than the latter, and that it does not slavishly follow existing practice, although there are aspects of current practice that it would approve. Repeatedly in different contexts, considered from different viewpoints, we saw that the behaviour recommended by the values stood up well to their application to consumer bodies and bankers. The results of the testing support our additional claim that the model would be understandable to all sides of the industry. In the remainder of this chapter we present the model to show its usefulness and limitations illustrating how it could be applied to inform the decision-taking process so that decisions would be ethically justifiable.

6.2 The Stewardship Debt Management Model

The model is presented in *Figure 6.1* to give a clear overview. Brief explanatory notes to make the diagram more intelligible follow this. The specific values included are then considered in detail, in the order in which they appear on the diagram, and illustrations are given of how they can facilitate ethical reasoning. In *Figure 6.1* the key stewardship values are shown in capital letters, namely, entrustedness, matching, effectiveness, development, accountability and fairness. The essence of stewardship is *entrustedness*, which narrowly construed is applied to resources and in its broad sense includes social considerations. The term *matching* is used to identify the necessary objective assessment by quantitative and qualitative measures, which is part of a reasoned response to resource management in an environment of scarcity. *Chapter 2* demonstrated the necessity of matching costs and benefits so that reasoned ethical choices about resources can be made. Matching appears twice in the diagram. In the first instance, as applicable to bankers and in the second, applied to borrowers. In the case of the banker the linkage of shareholders and depositors to operational lending decisions demands managerial *effectiveness*. This in turn requires relevant managerial objectives, which determine the nature of the measurement required in loan appraisal. This latter involves more

FIGURE 6.1 THE STEWARDSHIP OPERATIONAL MODEL

I	III
Shareholders	*Depositors*

ENTRUSTEDNESS*

|

MATCHING* (A) *BANKERS*

|

EFFECTIVENESS*

|

Objectives
Safety of funds
Cash Inflows
Remuneration

|

Loan Appraisal
including
Borrower Effectiveness

|

MATCHING* (B) *BORROWERS*
Information quality, relevance, and accuracy

II
Borrowers

Personal Borrowers		*Business Borrowers*
Solvency		Net Worth
Surplus income		Profitability
Self-assessment	------------------------------------	Personal factors
Expectations	------------------------------------	Economic factors
Strategic soundness	------------------------------------	DEVELOPMENT*
Financial control		Liquidity
Safety		Security

ACCOUNTABILITY*

|

I	II	III	IV
Shareholders	*Borrowers*	*Depositors*	*Community*

**FAIRNESS*
Supportiveness, Social Costs and IV. *Community* Norms & Expectations

than financial information. The centre of the diagram shows the interaction between lenders and borrowers, where the former must test, by *matching*, the effectiveness of the latter. It is possible to incorporate both personal and business borrowers within the model. As was explained in *Chapter 3* and *Chapter 4*, the appraisal of personal and business borrowing involves a generic stewardship model, but the appraisal of business propositions is more complex. The centre of the diagram shows, however, that both types of appraisal can be accommodated on the same matrix. *Development,* as we have seen in previous chapters, has a narrow *economic* sense and a broader *social* sense. *Accountability* is logically at the feedback point in the model. It would have made the diagram excessively complex to seek to incorporate arrows to each stakeholder but each stakeholder has been identified by a Roman numeral, in the diagram and under the heading *Accountability*. *Fairness* has been appended at the bottom, not because it is of less importance, but because it must inform reflection regarding *each* of the other values. This has been indicated by the asterisks, which draw attention to fairness. Fairness is not to be viewed as some filter at the end of the process. This can be illustrated, for example, by the need to apply fairness to the measurement process. To be fair to the banker, the borrower must provide accurate data, and to be fair to the borrower the banker must weigh the different components of the measurable data in the light of the demands of fairness. This is of particular relevance in banking where stakeholders are of *unequal* financial expertise. It is not that fairness changes the objective statement of the measurable data. It may, however, influence the data that is gathered and the process through which it is analyzed. These points arose with respect to credit referencing and credit scoring.

In the remainder of this chapter the contribution to the decision taking process of the various values will be considered with reference to borrowers and lenders.

6.2.1. ENTRUSTEDNESS

There are three relevant situations of entrustedness: shareholder-banker, depositor-banker and banker-borrower. For bankers, *entrustedness*, in its narrow sense of accountable control of resources, involves being charged with adding value to the shareholders' funds. In addition, depositors entrust funds to the banker for safekeeping, a modest return and repayment on demand or after due notice. In the broad sense, entrustedness involves shareholder and depositor expectations that bankers will go about their business in a socially responsible way. We do not entrust our funds to loan sharks because we do not like the way in which they would use them. In some cases stakeholder expectations of banks might be very specific, for example, when they do not expect funds to be lent to firms that would pollute the environment. In order to meet stakeholder expectations bankers must:

(a) safe-guard assets committed to their control,
(b) utilize the assets in a remunerative way,
(c) meet depositors' withdrawals on demand or after due notice,
(d) fulfil the social responsibility commitments of the bank,
(e) credit customers' accounts with their interest entitlement, and
(f) pay out cash dividends to shareholders and enhance the asset value.[1]

Bankers obtain remuneration on their assets by entrusting funds to borrowers. Regarding the narrow sense of entrustedness, bankers expect that borrowers will use these funds in a responsible way so that interest and repayments can be paid according to the agreed schedule. The broader aspect of entrustedness requires some reinforcement. Implied in the entrusting of funds to borrowers is the expectation that they will not use them for socially unacceptable purposes but this needs to be more clearly articulated. It could relate to the *kind* of products in a business loan. It could relate to levels of family care in the case of personal loans. Bankers, for example, should not condone ostentatious consumption related to child neglect.

6.2.2. MATCHING (A) BANKERS

In the technical texts, the banker's *matching* is related to both guardianship and returns. The former is addressed by maintaining an appropriate balance sheet asset structure, by matching assets and liabilities, so that there is an appropriate distribution of short and long-term assets. Return on investment is addressed by using the funds in a way that gives an interest return to the bank. The ratio of income to loaned funds determines the remunerative attractiveness of particular assets to the bank. There is also an issue of matching repayments and loans. Adequate bank liquidity necessitates timely cash inflows and this involves regular repayments by borrowers. Hard core borrowing, without significant repayments, even if secured, is incompatible with the bank's need to maintain regular inward cash flows. Regulatory frameworks involve monitoring of the matching of capital/total assets or other solvency ratios.

6.2.3. EFFECTIVENESS

Banker *effectiveness* is closely related to entrustedness because the situation described above under *Section 6.2.1* determines the bankers' objectives. We might summarize the operational focus as *remunerative lending within the constraints of asset liquidity and asset safety*. The three relevant objectives appear in *Figure 6.1* as *Safety of funds*, *Cash Inflows* and *Remuneration*. Operationally banker effectiveness means meeting these objectives, which, in particular cases, require the testing of borrower effectiveness, because, if borrowers fail, the bank incurs bad debts. Two dilemmas arise in connection with the three objectives. Firstly, as soon as funds are lent, their safety cannot be guaranteed, except in the case of gilt-edged securities, which are not very remunerative. Secondly, the most remunerative use of funds is to lend to consumers and businesses, but such lending is not liquid. We will consider the way in which our stewardship values can inform operational decisions by looking at the example of risk management in connection with this dilemma.

Stewardship Values and Ethical Risk Management

The stewardship model enables us to address the banker's risk management in an ethically sensitive way. Bankers operate with a risk-return trade off within a liquidity constraint and face three major risks relative to the specified *objectives*:

Risk A: Safety of funds failure.
Loss of asset value through borrower default.
Risk B: *Cash inflow* insufficient.
Inadequate liquidity to meet withdrawals.
Risk C: *Remuneration* shortfall.
Insufficient to fulfil stakeholder expectations.

How do stewardship values help? Consider narrow *entrustedness*. Depositors who are risk averse entrust bankers' main loan funds. This requires that the minimization of *Risk A* be the operational priority, because depositors prefer eventual repayment *in full*, even if the bank defaulted regarding repayment on demand, as a result of an illiquid position. Bankers must prioritize the management of *Risk B* in relation to *Risk* C because inadequate liquidity will put the bank or its customers out of business more quickly than inadequate returns. Such prioritization indicates that the banker should be conservative in his operational application of the risk-return trade off and that risk minimization should be prioritized before remuneration. This accounts for the order of our objectives in *Figure 6.1*.

This line of argument is relevant to broad *entrustedness* which relates to social responsibilities. The criticism is sometimes heard that it is socially detrimental that bankers are not sufficiently adventurous in their lending. It is assumed that, if they were, more business activity would be generated with beneficial outcomes for employment. But, as we have seen from the foregoing argument, bankers are *not* in the same position of entrustedness as commercial executives as opportunity seekers, because most of their funds come from risk-averse providers. It is not good ethics to expect banks to go contrary to the requisite *matching* of assets and liabilities to engage in risky lending to satisfy social objectives such as increased employment.

Risk A is obviously relevant to the balance between secured and unsecured lending, and this balance is obviously relevant to *Risk C.* In competitive environments there is the opportunity to improve remuneration by granting marginal propositions and ensuring that *Risk* A is covered by adequate security. Although, the standard textbook position, as seen earlier in Rouse [2], would be that banks should not rely on security to decide the outcome of loan requests, there is little objection on the grounds of *effectiveness*. It might be argued that too many default situations with the bank calling in security will generate adverse publicity, and this would need to be considered; but a strategy based upon secured marginal lending could, from the banker's point of view, be remunerative and safe. Such an advantage raises, therefore, the need for other values to secure an appropriate balance in the banker-customer relationship and in relation to the social costs of bad lending. In applying the stewardship *concept* we are applying a *set* of values and not isolating single values to apply them without reference to the others. Lending that would improve unemployment in the short-term could meet the social aspect of *development* but we would still be faced with the banker's accountability to stakeholders, including borrowers. Previous chapters have shown that what might be effective from the bank's point of view must be modified by what is fair from the borrower's standpoint. We have said that *fairness* must not be isolated from the application of the other values. Under fairness in *Figure 6.1,* we have identified *supportiveness* and *social* costs which require: (a) that banks resist pressures to more adventurous lending where it would be to the detriment of borrower solvency and (b) that the safety and remuneration of secured debt should not over-ride borrower protection measured by *borrower* effectiveness. We saw in *Chapter 5* that this is especially relevant to mortgages. The latter may be good for builders and their employees but disastrous to ineffective borrowers. The importance of checking the latter is flagged in *Figure 6.1* as *Loan Appraisal including Borrower Effectiveness*. This effectiveness is tested by matching to which we now turn.

6.2.4. MATCHING (B) BORROWERS

Matching (B) is at the centre of *Figure 6.1* reflecting its centrality in ethical decision taking. We saw this in *Chapter 2* when considering stewardship in the environmental context. Ethics must not be isolated from matching and vice versa. We will consider the significance of matching, then look at key factors and practical approaches.

Significance

The strategic importance of matching for operational stewardship has been indicated in previous chapters. We saw in *Chapter 2* that any attempt to deploy and control resources in an effective way necessarily involves some form of matching so that inputs and outputs can be compared to determine whether or not a net benefit arises. We saw in *Chapter 3* that this is reflected in personal borrowing through the need to match acquisitions with income availability, present or future, to avoid financial default with its personal or social costs. It is easy to regard matching as a merely technical matter to be tailored to meet the bank's needs. We drew attention to this problem in relation to credit scoring but the importance of reasonably accurate measurements in personal borrowing was underlined by empirical research referred to in *Chapter 3*. Factors contributing to financial difficulties fall into two main categories. There are factors over which the borrower has little control such as loss of employment, increase in family size or health problems. There are also factors that arise from the way in which borrowers control their finances such as over-commitment and multiple debt. Regarding both categories the nature and quality of the matching used are of major importance. Quantitative "what if" analysis can clarify the degree of risk involved in relation to the uncertainties of life with a view to avoiding over-commitment. The banker's customary role addressed these needs to some extent, because, in talking through a proposition with the customer, such issues would be highlighted. The move to technology-based personal lending has taken out the banker-customer interaction that served as a check on an over-enthusiastic use of easily available credit, vigorously promoted by persuasive advertising. Our case studies have shown that this has not just affected the financially unsophisticated. In consequence of interest rate changes, equity erosion, interest rate relief alterations and structural changes in the provision of mortgage finance, many professionals have defaulted. This reveals that bankers should not assume that an advisory role is expendable with such customers. This role should not be divorced from the banker's scrutiny of the customer's financial position as revealed by financial matching. We have seen that credit scoring allows bankers set their risk to order, making bad debts an acceptable proportion of total commitments. The ethical demands impose a twofold problem for bankers. They must assess what they need to work through with their customers so that the latter will be able to effectively order their financial affairs to repay the entrusted funds. Bankers must further decide how to approach the scrutiny and interaction with the customer without increasing lending costs. Our stewardship model helps with these tasks.

Key Factors

Our analysis of the measurement requirements entering into personal borrowing was conducted in the light of empirical findings concerning consumer borrowing. In *Chapter 3* we looked at consumer borrowing from a stewardship perspective and saw that there were seven key factors.

(1) A realistic matching of assets and indebtedness.
(2) Matching personal income and expenditure to determine the surplus.
(3) An informed self-assessment by the borrower.
(4) Realism of borrower expectations and the consequent targets set.
(5) The strategic soundness of the borrowing.
(6) Arrangements for financial control should be considered.
(7) Safety is an important consideration for bankers and borrowers.

Practical Approaches

These components of the measurement aspect of our stewardship model are necessary to ensure responsible borrowing and avoid default and need to be addressed. How can this to be accomplished? Bankers will not want to return to individual appraisal for all personal lending and the increased interview costs would not be popular with borrowers. If a technological answer is the only course, the diagnostic questions should be extended to include a scrutiny of the factors identified above. Results could then be incorporated in a feedback to the customer highlighting specific areas of risk or weaknesses of financial management. Different levels of banker response could be built in for different diagnostic results. Possibilities for a scaled response include: a financial planning pack, a financial monitoring pack, or the offer of personal advice at the local branch if desired. Bankers need in some way to build in diagnosis helpful to individual borrowers ensuring a better balance between profitability and customer needs.

6.2.5. DEVELOPMENT

The Nature of Development

The meaning of the term development, as involving economic growth *and* social responsibility, is the same for lenders and borrowers, but the implications differ because of different circumstances and relationships. For bankers development involves return on investment for shareholders, interest return to depositors and socially responsible operation. Being a financial intermediary, a bank serves as a channel of stakeholder funds into economic development through lending. The banker depends upon economic development in *other* businesses. *Business* purpose is not an essential criterion for borrowing, but participation in the total economic development process is. With respect to a borrower's business development, bank stakeholders have first claim on the economic returns so the banker must ensure the effective use of loaned funds in appropriate economic development to ensure repayment. Concerning broader social responsibility the banker must be satisfied that the economic returns of the development are not outweighed by social disbenefits. Employed borrowers contribute to economic development through their work and secure income to effect repayment. Personal loans need not be directly related to the development activity that provides the income but in some cases personal loans contribute directly to the economic development and the banker should be interested in its nature and continuance. The broad sense of development is also relevant and personal borrowers should face up to the implications of social development in terms of total family welfare. Our stewardship model would disapprove of a family running up debts for a parental holiday at the expense of the children's standard of living. Borrowers ought to consider the relevance of development as applied to lenders and the implications of the total context of development for themselves in both its narrow and broad dimensions.

The Importance of Development

In previous chapters, we saw that development should have a bearing on business and consumer objectives where borrowed funds are used. Traditionally the loan purpose has been an important criterion in the appraisal, especially for business finance. Speculative purposes should be viewed unfavourably compared to investment purposes. Financing business equipment and stocks have been viewed as self-liquidating, because the revenue provides income to meet repayments. With respect to consumer borrowing our model favours an asset acquisition purpose such as consumer durables, property extensions, or financing home ownership. There is now something of a paradox with respect to the importance of development. On the one hand, it has assumed a greater importance in relation to environmental matters. Banks have accepted that the evaluation of proposals can no longer be confined to internal costs. Financial analysis of a borrowing firm's performance and position, as outlined in *Chapter 4* is still considered important, but so are the external costs of environmental pollution. It is accepted that the measurement of genuine development embraces *more* than the bottom line profit figure resulting from matching internal costs and benefits. The net gain must not be offset by a higher cost environmental degradation. Despite this recognition of the principle of sustainable development with respect to the environment, lenders have, on the other hand, diminished their emphasis upon development with respect to personal borrowing. It is now no longer uncommon for credit promotion materials to advertise loans for *any* purpose, so that borrowers are not constrained as to how they spend the entrusted funds, be it for a fitted kitchen, furniture, holiday or whatever else. In addition mortgage deals are being advertised, at the time of writing, offering substantial cash-back incentives with a completely open consumer choice as to how the money is spent. Such deals have already attracted official concern because of the potential danger to lenders' capital strength in the event of savers making withdrawals or interest rates rising but they also emphasize the drift from developmental purpose. This paradox illustrates a major problem with respect to ethical debt management. Too often values are adopted piecemeal, and in relation to a specific context, as a reaction to public pressure, and, in the case of the environment, potential lender liabilities. But the same value ends up neglected with respect to other contexts. This situation shows the importance of a *set of core values* that will be consistently applied by financial institutions in the different contexts relevant to their lending operations. As the three previous chapters have shown, the values derived from the stewardship concept provide such a core, capable of illuminating moral issues in operational practice and of recommending relevant actions to promote ethical debt management, which set development in its full context.

6.2.6. ACCOUNTABILITY

Accountability enters into mutual responsibility and trust. We have indicated the importance of accountability in *Figure 6.1* by linkages back to shareholders, borrowers, depositors, and the community. Accountability is exercised through various mechanisms. Managerial accountability to shareholders was discussed in *Chapter 2* in relation to governance. The main accountability mechanisms are annual audits, annual reports and accounts and annual general meetings. Accountability to customers, whether borrowers or depositors, is centered in contractual responsibility. The customer's route to holding the bank accountable for default in the banker-customer relationship would be to resort to the courts and to seek damages for breach of contract. The banking ombudsman provides a cheaper and less risky alternative for settling disputes where customers consider that bankers have failed in service provision. The main areas of complaint received are lending, mortgages and charges. The banker-borrower accountability is two-way and we have several times stressed the importance of adequate banker monitoring of loans. Accountability to the community is less structured except for legal responsibilities. The expectations of the community are expressed in various legislative provisions concerning the conduct of business. These include the requirements to comply with the various companies acts and banking acts and the regulatory structures that they create. They include from time to time credit controls and credit advertising. There are also less structured mechanisms. Media coverage of particular actions of public concern may result in the exposure of what is generally considered to be bad practice and this adverse publicity may compel the offending bank to change its ways. More organized reactions such as consumer boycotts may result from a lack of managerial responsiveness to public concerns. One route to formalizing accountability to the community is through a social audit and certain banks have adopted this to some extent. Our stewardship model would encourage financial institutions to do more to formalize accountability to stakeholder groups on the basis of the core of values to establish accountability. We give below some specific illustrative applications relative to shareholders, depositors and borrowers.

Stewardship and Socially Responsible Value-enhancement

In a market economy asset protection and value enhancement are legitimate shareholder expectations which our stewardship model does not undermine. But there is the broader entrustedness to account for and our model sheds light on managerial accountability with respect to profit maximization verses social responsibility. In the light of the relationship between accountability and entrustedness the conflict between profit and philanthropy can be seen to be resolvable. There are social responsibilities that all bankers should recognize but also a spectrum of acceptable positions beyond these. At one end are, what we shall call, *Category A* firms, which, while endeavouring to be fair to employees and customers in all circumstances, prioritize the profit objective. At the other end of the spectrum are, what we shall call *Category E* firms, which see their role as involving a very heavy social commitment through a variety of community involvement schemes. Viewing accountability in the light of entrustedness we can see that this need not give rise to conflict for bankers. The issue is not whether *Category A* is right and *Category E* wrong or vice versa; rather the issues for managers are:

(a) the terms upon which assets have been entrusted to them,
(b) compliance with those terms, and
(c) accurate disclosure of performance.

It is not for business ethicists to preclude investors from entrusting funds to bankers at either end of the spectrum or somewhere in between, but it is proper to maintain that investors should be clearly told how the funds will be used, and have it confirmed to them that the funds have been used in that way. We saw in *Chapter 2*, that the *Objects Clause* of a *Memorandum of Association* is too general to communicate the detail of a bank's commitments. Additionally, changing management policy regarding community involvement is not a contractual matter. There are ways of improving this situation. The *Registrar of Companies* could, for example, define separate categories on the spectrum of social responsibility commitment. Each category: A, B, C, D and E could be defined in terms of the percentage of surplus revenue to be utilized for community involvement, providing a quantitative criterion, and the category could also be defined in some qualitative way. Each registered company could then register its positioning in terms of one of the specified categories. In this way there would be a clear definition of each company's contractual position. Under such a system a bank's management, like the management of any other company, would have a clear marker of investor expectations to which to work and in relation to which they could report their effectiveness. The issue at the accountability reporting stage would be that there was a fair disclosure of management performance with respect to social responsibility commitments. Investors would be in a position to determine whether or not a particular bank was in a category in which they wanted to invest, and whether or not the management was adequately discharging their task. What would be unacceptable from a stewardship perspective would be for the managers to appropriate funds to social responsibility objectives contrary to the expectations of shareholders. In support of this position it may be said that best practice in responsible firms provides parallels to some of these suggestions as can be seen in the ethical positioning of banks such as the TSB and the Coop Bank.

Stewardship and Depositor Protection.

Bankers are morally accountable to depositors because entrustedness embraces, not only the situation where parties are charged with the care and use of assets upon very specific terms, but also the situation where resources are committed to the care of a third party. The banker-depositor relationship is of this latter kind. Depositors entrust funds to the bank for safekeeping; and repayment on demand and in the expectation of a return by way of interest. Bankers are thus guardians of the funds but it is understood that, they must also exercise a stewardship of them to provide depositors with a return on savings and withdrawals when required. As bankers seek profits they must lend in a way that is consistent with the safety of the funds. Depositors' funds dwarf shareholder funds, so that the management's responsibility to depositors in quantitative terms is *greater* than their responsibility to their shareholders. This is not adequately reflected in accountability mechanisms and depositors are left relying on regulatory capital adequacy ratios. Depositors are more risk averse than bank shareholders whose investments are usually diversified, whereas depositors may have all their savings in a single bank. Our model suggests that banks should do more to account to depositors to demonstrate a sufficiently cautious lending policy.

Stewardship and Borrower Monitoring

Our stewardship model has shown that bankers should not be lax about borrower accountability. Entrustedness involving the commitment of resources to a third party arises in lending. As the depositor commits funds to the banker, the latter commits funds to the borrower in the expectation that repayment will be made and charges met. This situation is so similar to the depositor-banker relationship that it would be remarkable to suggest that the banker's expectations of his borrower should be different from the depositor's expectations of his bank in this respect. The bank cannot regard deposited funds as absolutely its own, but must account for them to the depositor, and the borrowers should not regard loaned funds as absolutely their own, but an accountability to the bank should be recognized. In both cases there is a responsibility to return the funds to the owner as required under the terms of the arrangement. In both cases there is an undertaking to pay interest. In fact it is the same funds and the borrower is accountable to the banker as the banker is accountable to the depositor. This has some relevance to the subject of default. When bankers have funds entrusted to them, they are expected to be accountable and not default on their commitments to depositors. When bankers entrust funds to borrowers who default, they are often expected to extend debt forgiveness. There is no consistency in such an approach as the same kind of entrustedness and accountability applies to both situations. Our model expects banks to repay their depositors and expects borrowers to repay their lending bankers; but society often discriminates against the banks in this respect. A balanced approach, as we have seen in *Chapter 5*, would limit debt forgiveness to cases where the bank is, in part, responsible for the default. The objective bases upon which bankers could determine the extent of debt remission to be granted would be (a) the extent to which professional failure in granting the loan had contributed to the default, (b) the saving to the bank's depositors and shareholders, by writing off the debt, and (c) any bank failure of monitoring. More transparency about this latter point would strengthen accountability mechanisms concerning monitoring loan funds.

6.2.7. FAIRNESS

We have said that issues of *fairness* arise in the application of each of the values and for this reason our model requires its consideration throughout. In *Chapter 2* we saw that the "rules of the game" in the case of banking requires a breadth of interpretation because within the banker-customer relationship the two parties are often of unequal standing.

This is particularly the case with personal borrowing and the self-employed where the banker has superior financial expertise. Bankers can thus look after their self-interest in ways that prove detrimental to customers. Bankers can cover the default risk by taking security and to meet targets it is an option to improve remuneration by granting marginal propositions, especially when there are surplus funds for lending. Our model does not regard this as fair to customers unless they fully understand the situation. A further conflict identified in *Chapter 4* related to credit scoring and referencing because systems appraisal is effective from the banker's point of view reducing appraisal costs at defined risk levels. With respect to default risks, systems appraisal provides a tool to control bad debts and balance risk and remuneration. Our concern is that it is not necessarily *fair* banking. The system enables bankers to allow for non-payers without losing expected returns, but it does not disclose the individual weaknesses of borrower stewardship that cause default when the uncertainties of life produce a crisis.

When fairness is considered, our model expects that the financial expertise of the banker should be exercised to reduce borrower risk as well as lender risk. This can only be accomplished by reinforcing our stewardship values to minimize borrower exposure to financial difficulty. Such a focus would reduce the emphasis upon the probability of repayment, and increase the emphasis upon total commitment and the realism of the customer's expectations in the light of resources, particularly existing assets and expected income. Bankers should be satisfied that customers are aware of risks involved and have thought through personal solvency, the realism of personal objectives in the light of financial resources, and the security of the intended source of repayment allowing for unexpected events. The strength of acquisitiveness, manipulative advertising and over-enthusiastic use of easily obtained credit by poorly informed and financially unsophisticated consumers, requires bankers to satisfy themselves about borrower effectiveness. This requires a blend of scrutiny, advice and tact that will produce the desired outcome in a way that is neither oppressive, nor unduly intrusive and which takes place within a relationship of confidentiality. Thus when effectiveness and fairness are both taken into consideration, it is seen to be important that the banker thoroughly tests *borrower* effectiveness and this requires appropriate tools for the task.

6.2.8. LIMITATIONS OF THE MODEL

This section has shown that stewardship values are capable of application to the operational lending situation and do stimulate ethical thinking and recommend improvements in practice but there are limits. In this section we look at the different values in order to show some of the limitations that must be born in mind.

Entrustedness
We saw at the outset of our analysis of the stewardship concept in *Chapter 2* that entrustedness is at the heart of business relationships but that, as encountered in the accountant's world, the application of the concept can be merely contractual. Stewardship entrustedness will be severely limited if its moral force is not recognized. We saw that agency theory looks at the relationship between managers and shareholders in terms of principal and agent and the agency-information approach seeks to analyze the relationship in order to establish an *efficient contractual basis*. The legalistic view of stewardship, still prominent in accounting practice, also focuses attention upon the contractual performance of duties involving compliance with company objects and accounting standards. While the importance of various formal prescriptions to secure minimum standards of performance must be recognized we have already made clear the limitations of codes from a moral point of view. Entrustedness as a value will be likewise limited if it is interpreted in a narrow legalistic way. We commenced our analysis of the stewardship concept in *Chapter 2* by showing that, not only can its significance be extended from the legalistic to the economic, but from both of these to the moral. The force of our set of values as an agent for change lies in the adoption of the concept as a source of *moral* values to be applied in a strategic not formal way.

Matching
Matching has been included as a key feature of ethical debt management. The model shows the strategic place that matching must occupy in ethical lending and borrowing. Our analysis of the stewardship concept showed that measurement must be based upon matching costs and benefits and underlines the importance of reasonable accuracy in measurements for the purposes of matching. The model itself does not shed light upon how to attain technical accuracy of measurement of costs or benefits. We indicated this in *Chapter 2* where we saw the difficulties involved with social cost calculations. Indeed, it

would be difficult to apply the model in a context where there are no meaningful market prices, for, although stewardship can operate in a small uncomplicated community without market pricing, there is no way of visualizing how the model could operate with any degree of objectivity in a complex economy, without market values as points of reference for rational decisions. It must be concluded that issues of valuation are independent of the model and that, although our stewardship analysis has shown that matching is integral to stewardship in ethical business activity and is essential to the application of the stewardship lending model, the avenues for deriving reasonably accurate valuations for costs and benefits must be sought elsewhere.

Effectiveness

Effectiveness is important for lenders and borrowers if the demands of entrustedness are to be satisfied. Effectiveness of action necessitates appropriateness of decision. It is not essential for the decision to be the best decision for it to be relevant and useful. However, better decision taking results in greater effectiveness, other things remaining equal. The stewardship model presented here is not of that rationalistic kind which makes the decision on behalf of the human being by a process of rationalistic prediction. The stewardship lending model serves to illuminate the ethical considerations involved so that the decision is, in value terms, a better decision. It cannot make the decision on behalf of the manager but it can assist the manager to review important ethical considerations so that there is an improved final decision about what needs to be done. Thus, by way of illustration, reference can be made to the discussion above, where the point was made that the banker, having decided to make a personal loan on the basis of a particular kind of risk assessment, could extend the bank's contribution to the borrower's understanding of the situation by providing relevant information and advice to help the borrower to assess and cover the *borrower* risks involved. The stewardship lending model indicates that this would be a more responsible approach and that the final decision would be a better one if it included this particular course of action. This is, however, but one example. The purpose of the model is to provide a framework within which the manager can think about the value issues of his profession and a particular loan application with a view to improved value-sensitivity in the decision taking. What the model cannot do is replace the manager's own thinking through of the particular case, make decisions on behalf of the manager or decide how best to organize responsive actions.

Development
Development focuses attention upon objectives and desirable targets. Stewardship emphasizes that development is important and provides some indicators as to what is and is not development in relation to resources. Ordinarily development can be identified by an increase in available resources and resultant products. We must not, however, expect too much of the Stewardship Model as though it were, in itself, a complete model for moral decision. There are indeed certain questions related to outputs that it can shed light upon. So, for example, if we view our own bodies as necessarily involved in our stewardship, then what damages the body would be viewed negatively from the standpoint of stewardship development. This would lead us to rule out, say, drug taking and production as beneficial activities. Stewardship can also be a powerful moral guide to domestic expenditure. Many Christians, for example, test personal expenditure by the question, Is this good stewardship? Thus, on both the producer side and the consumer side, stewardship can be serviceable in guiding action with respect to the allocation of resources. No claim is being made, however, that stewardship can be a *complete* guide in and of itself. As the marketplace is, as we have seen in *Chapter 1,* a serviceable tool, so stewardship is a serviceable concept, but neither the marketplace nor the stewardship concept should be elevated to a position of being supreme, exclusive and all sufficient in determining morality. Both presuppose a broader ethical context, within which they serve their respective purposes. We would not, for example, decide between saving the life of a child or saving the life of a grandparent purely upon the grounds of which might be the best stewardship.

Accountability
Procedures for accountability constitute a research area of their own. The stewardship model is suggestive of the kinds of issues that would need to be included and the general standards that might reasonably be expected by the different stakeholders. The model is, however, a general model and does not prescribe the particular routes by which managerial accountability to the different constituencies can be achieved.

6.2.9. CONCLUSION

As we have seen in *Chapter 5*, the operational usefulness of the stewardship concept receives empirical support from the success of the *CCCS* because the underlying ethic of the latter implies entrustedness, by upholding a repayment ethic; incorporates supportiveness in providing advice; aims at effectiveness of repayment through debt management plans, and focuses on matching income and expenditure to implement repayment. The positive response of creditors to financing the money advice by voluntary contributions related to repayments supports our emphasis on accountability. We can conclude that the stewardship approach developed in this work would serve as a suitable basis for a national strategy and could be supported by financial institutions, consumer organizations and advice agencies.

Our overall conclusion is that the testing in *Chapters 3* to *5* supports our initial proposition that the values derived from the stewardship concept provide an understandable moral core, capable of acceptance by borrowers, lenders and advice agencies. The preceding chapters have shown that the values, when applied together, shed light upon operational decisions and promote ethically sensitive debt management practice.

Notes to Chapter 6

1. For technical details see Jeff Madura, *Financial Markets and Institutions,* pp.421-441, 475-507.
2. C. Nicholas Rouse, *Bankers' Lending Techniques,* p.7.

Bibliography

Adams, Robert M. 'Religious Ethics in a Pluralistic Society'. In (eds.) Outka, Gene and Reeder, John P., Jr. *Prospects for a Common Morality.* Princeton, New Jersey: Princeton University Press, 1993.

Andrews, David. 'Dealing with debt'. p.41. *Credit Management*, July 1989.

Andrews, Kenneth R. 'Can the Best Corporations be Made Moral?' *Harvard Business Review*, May-June, 1973. Reprinted in *Ethics for Executives: Part II*, USA: Harvard Business School, 1977.

Atherton, John. 'New Occasions Teach New Duties? 9. The Individual and the Organization.' pp. 356-362. *The Expository Times*, September 1994.

Barcena, Alicia and Payne, Noel D. 'Stewardship and Sustainable Development.' pp.33-36. *National Forum,* vol.75 no.1.1995.

Barty-King, Hugh. *The Worst Poverty: A History of Debt and Debtors.* Stroud: Alan Sutton Publishing Ltd., 1991.

Battiscombe, Georgina. *Shaftesbury: The Great Reformer 1801-1885.* Boston: Houghton Mifflin Company, 1975.

Beauchamp, Tom L. and Bowie, Norman E., (eds.) *Ethical Theory and Business.* Englewood Cliffs, New Jersey: Prentice-Hall, Inc., 1979.

Beauchamp, Tom L. and Bowie, Norman E., (eds.) *Ethical Theory and Business*, Fourth Edition. Englewood Cliffs, New Jersey: Prentice-Hall, Inc., 1993.

Berry, A.J., Faulkner, S., Hughes, M. and Jarvis, R. 'Financial Information, the Banker and the Small Business'. pp. 131-150. *British Accounting Review,* 25, 1993.

Berthoud, Richard. *Credit, Debt and Poverty.* London: HMSO, 1989.

Berthoud, Richard and Kempson, Elaine. *Credit and Debt: The PSI Report.* London: Policy Studies Institute, 1992.

Beutler, I.F., Burr, W.R., Bahr, K.S., and Herrin, D.A. 'The Family Realm: Theoretical Contributions for Understanding its Consequences.' pp.805-816. *Journal of Marriage and the Family*, 51(3), 1989.

Block, Peter. *Stewardship: Choosing Service Over Self-Interest.* San Francisco: Berrett-Koehler Publishers, 1993.

Bolton, J.E.(Chairman). *Small Firms: Report of the Committee of Inquiry on Small Firms.* London: HMSO, 1971.

Borrie, Gordon. 'Law and Morality in the Marketplace'. In (eds.) Prindl, Andreas R. and Prodhan, Bimal. *Ethical Conflicts in Finance.* Oxford: Blackwell, 1994.

Bowie, Norman E. 'Business Ethics as a Discipline: The Search for Legitimacy'. In (ed.) R. Edward Freeman, *Business Ethics: The State of the Art.* Oxford: Oxford University Press, 1991.

Bowie, Norman E. and Freeman, R. Edward (eds.) *Ethics and Agency Theory: An Introduction.* Oxford: Oxford University Press, 1992.

Brenner, Steven N. and Molander, Earl A. 'Is the ethics of business changing?' *Harvard Business Review*, January-February, 1977. Reprinted in *Ethics for Executives: Part II.* USA: Harvard Business School, 1977.

Briggs, Asa. *Victorian People: Some Reassessments of People, Institutions, Ideas and*

Events, 1851-1867. London: Odhams Press Ltd., 1954.

Bromwich, Michael. *Financial Reporting, Information and Capital Markets.* London: Pitman, 1992.

Brown, Colin (ed.) *New Testament Theology, Volume 2.* Exeter: Paternoster Press, 1976.

Brown, Halina S., Derr, Patrick, Renn, Ortwin and White, Allen L. *Corporate Environmentalism in a Global Economy.* London: Quorum Books, 1993.

Browne, Andrew Howe. 'The Banks and Personal Credit.' pp.17-19. *Banking World,* January 1991.

Bush, President George. 'A New Era of Environmental Stewardship.' pp.2-3. *EPA Journal,* September-October, 1990.

Buttrick, George A., (ed.) *The Interpreter's Dictionary of the Bible: An Illustrated Encyclopedia.* Nashville: Abingdon, 1980.

Cadbury, Adrian. 'Ethical Managers Make Their Own Rules'. In (eds.) Prindl, Andreas R. and Prodhan, Bimal. *Ethical Conflicts in Finance.* Oxford: Blackwell, 1994.

Calvin, John. *Calvin's Commentaries: 1. The Pentateuch.* Grand Rapids: AP&A, Undated.

Cannon, Tom. *Corporate Responsibility.* London: Pitman/Financial Times, 1992.

Carmichael, Sheena and Drummond, John. *Good Business: A Guide to Corporate Responsibility and Business Ethics.* London: Business Books, 1989.

Carr, S. *Conservation on Farms: Conflicting Attitudes, Social Pressures and Behaviour.* Open University: Ph.D., 1988.

Central Statistical Office. *Annual Abstract of Statistics.* London: HMSO, 1996.

Certo, Samuel C. and Peter, J. Paul. *Strategic Management: Concepts and Applications.* McGraw-Hill Inc., 1991.

Checkley, Keith. *Finance For Business: A Guide for the Lending Banker.* London: Financial Training Publications Limited, 1988.

Chen, Rosita S. 'Social and Financial Stewardship.' pp. 535-543. *The Accounting Review.* July 1975.

Childress, James F. and MacQuarrie, John (eds.) *A New Dictionary of Christian Ethics.* London: SCM, 1986.

Church of England, *Something to Celebrate: Valuing Families in Church and Society. The Report of a Working Party of the Board for Social Responsibility.* London: Church House Publishing, 1995.

Common, Michael. *Environmental and Resource Economics: An Introduction.* London: Longman, 1988.

Congdon, Tim. *The Debt Threat: The Dangers of High Real Interest Rates for the World Economy.* London Blackwell, 1989.

Cowton, Christopher J. 'The Development of Ethical Investment Products'. In (eds.) Prindl, Andreas R. and Prodhan, Bimal. *Ethical Conflicts in Finance.* Oxford: Blackwell, 1994.

Credit Action. *Mind Over Money: Participants' Notes.* Cambridge: Credit Action, 1991.

Daft, Richard L. *Organization Theory and Design.* Third Edition. St. Paul: West

Publishing Company, 1989.
Davies, Alan. 'Too much income: too little saving.' p.42. *Banking World,* April, 1990.
Dees, J. Gregory. 'Principals, Agents and Ethics'. In (eds.) Bowie, Norman E. and Freeman, R. Edward. *Ethics and Agency Theory: An Introduction.* Oxford: Oxford University Press, 1992.
Donaldson, John. *Key Issues in Business Ethics.* London: Academic Press, 1989.
Donaldson, John. *Business Ethics: A European Casebook.* London: Academic Press, 1992.
Donaldson, Thomas. *The Ethics of International Business.* Oxford: Oxford University Press, 1989.
Doyle, E.P. *Law Relating to Banking.* Worcester: Northwick Publishers, 1981.
Drucker, Peter F. *Managing in Turbulent Times.* London: Pan Books, 1981.
Dyer, L.S. *A Practical Approach to Bank Lending.* London: The Chartered Institute of Bankers, 1987.
Elliott, Barry and Elliott, Jamie. *Financial Accounting & Reporting,* Second Edition. London: Prentice Hall, 1996.
Evan, William M. and Freeman, R. Edward. 'A Stakeholder Theory of the Modern Corporation: Kantian Capitalism.' In (eds.) Beauchamp, Tom L. and Bowie, Norman E. *Ethical Theory and Business*, Fourth Edition. Englewood Cliffs, New Jersey: Prentice-Hall, Inc., 1993
Fairholm, Gilbert W. *Values Leadership: Toward a New Philosophy of Leadership.* New York: Praeger, 1991.
Faith in the City. London: Church House Publishing, 1985.
Fleming, John E. 'Authorities in Business Ethics.' pp. 213-217. *Journal of Business Ethics,* no. 6, 1987.
Fleming, M.C. and Nellis, J.G. 'The Rise and Fall of House Prices: Causes, Consequences and Prospects.' pp.34-51. *National Westminster Bank Quarterly Review*, November, 1990.
Ford, Janet. *The Indebted Society: Credit and Default in the 1980s.* London: Routledge, 1988.
Frederick, William C., Post, James E., and Davis, Keith. *Business and Society: Corporate Strategy, Public Policy, Ethics,* Seventh Edition. New York: McGraw-Hill, Inc., 1992.
Freeman, R. Edward, (ed.) *Business Ethics: The State of the Art.* Oxford: Oxford University Press, 1991.
Friedman, Milton. 'The Social Responsibility of Business'. *Capitalism and Freedom*, pp.133-136, University of Chicago Press, 1962. Reprinted in (eds.) Beauchamp, Tom L. and Bowie, Norman E. *Ethical Theory and Business.* Englewood Cliffs, New Jersey: Prentice-Hall, Inc., 1979.
Friedman, Milton. *The Counter-Revolution in Monetary Theory: First Wincott Memorial Lecture.* London: Institute of Economic Affairs, 1970.
Friedman, Milton, in Gordon, Robert J. (ed.). *Milton Friedman's Monetary Framework: A Debate with His Critics.* Chicago: The University of Chicago Press, 1974.

Galbraith, John K. 'The Dependence Effect'. In (eds.) Beauchamp, Tom L. and Bowie, Norman E. *Ethical Theory and Business*. Englewood Cliffs, New Jersey: Prentice Hall, Inc., 1979.

Gardner, E. Clinton. *Justice and Christian Ethics.* Cambridge: Cambridge University Press, 1995.

Garnett, E.J. *Aspects of the Relationship between Protestant Ethics and Economic Activity in Mid-Victorian England.* Oxford: D.Phil., 1986.

Garrison, Sharon H. *The Financial Impact of Corporate Events on Corporate Stakeholders*. London: Quorum Books, 1990.

Geisler, Norman L. *Christian Ethics.* Leicester: Apollos, 1990.

George, Susan. *A Fate Worse than Debt.* London: Penguin, 1989.

George, Susan. *The Debt Boomerang: How Third World Debt Harms Us All.* London: Pluto Press, 1992.

George, Susan and Sabelli, Fabrizio. *Faith and Credit: The World Bank's Secular Empire.* San Francisco: Westview Press, 1994.

Gilbert, Daniel R. Jr. *The Twilight of Corporate Strategy: A Comparative Ethical Critique.* Oxford: Oxford University Press, 1992.

Gill, John and Johnson, Phil. *Research Methods for Managers.* London: Paul Chapman Publishing Ltd., 1991.

Gordon, Robert J. (ed.) *Milton Friedman's Monetary Framework: A Debate with His Critics.* Chigaco: University of Chicago Press, 1974.

Goudzwaard, Bob. *Capitalism and Progress: A Diagnosis of Western Society.* Grand Rapids: Eerdmans, 1979.

Goyder, George. *The Responsible Company.* Oxford: Basil Blackwell, 1961.

Gray, Rob, Bebbington, Jan and Walters, Diane. *Accounting for the Environment: The Greening of Accountancy, Part II.* Liverpool: Paul Chapman, 1993.

Griffiths, Brian. 'The Conservative Quadrilateral'. In (ed.) Chaplin, Jonathan, *When Christians Disagree: Politics and the Parties.* Liverpool: Inter-Varsity Press, 1992.

Guy, Mary E. *Ethical Decision Making in Everyday Work Situations.* London: Quorum Books, 1990.

Hanks, Roma S. and Sussman, Marvin B. 'Where Does Family End and Corporation Begin: The Consequences of Rapid Transformation.' In (eds.) Hanks, Roma S., and Sussman, Marvin B. *Corporations, Businesses, and Families.* London: The Haworth Press, 1990.

Hanna, Roberta. 'Families in Debt: The People and Their Money Problems etc.' In (ed.) Hartropp, Andrew, *Families in Debt.* Cambridge: Jubilee Centre Publications Ltd., 1988.

Hartley, Robert F. *Business Ethics: Violations of Public Trust,* New York: John Wiley & Sons, Inc., 1993.

Hartropp, Andrew. *The Debt Trap: The Impact of Credit on Families and Policy Proposals for the Future.* Cambridge: Jubilee Centre Publications Ltd., 1988.

Hartropp, Andrew (ed.); Hanna, Roberta; Jones, Simon; Lang, Raymond; Mills, Paul; and Schluter, Michael. *Families in Debt: The Nature, Causes and Effects of Debt Problems and Policy Proposals for Their Alleviation.* Cambridge: Jubilee Centre

Publications Ltd., 1988.
Hastings, James (ed.) *Encyclopaedia of Religion and Ethics, Vol. XII.* Edinburgh: T & T Clark, 1974.
Hayek, F.A. 'The Non Sequiter of the "Dependence Effect".' *Southern Economic Journal,* April, 1961. In (eds.) Beauchamp, Tom L. and Bowie, Norman E. *Ethical Theory and Business.* Englewood Cliffs, New Jersey: Prentice-Hall, Inc., 1979.
Hayek, F.A. *Denationalisation of Money: The Argument Refined,* Third Edition. London: Institute of Economic Affairs, 1990.
Heelas, Paul, 'Reforming the Self: Enterprise and the Characters of Thatcherism'. In (eds.) Keat, Russell and Abercrombie, Nicholas. *Enterprise Culture.* London: Routledge, 1991.
Hemming, Christine. *Business Success from Seizing the Environmental Initiative.* Letchworth: Technical Communication (Publishing) Ltd., Undated.
Henderson, Hazel. 'From Environmentalism to Ecophilosophy: Retooling Cultures for the Twenty-First Century.' In (eds.) Hoffman, Michael, Frederick, Robert, and Petry, Edward S. Jr. *The Corporation, Ethics and the Environment.* Westport: Quorum Books, 1990.
Henderson, Verne E. *What's Ethical in Business?* London: McGraw-Hill, Inc., 1992.
Henry, Matthew. *An Exposition of the Old and New Testament.* London: James Nisbet & Co. Limited, Undated.
Herbert, Alicia and Kempson, Elaine. *Water Debt and Disconnection, Research Results.* London: Policy Studies Institute, 1995.
Higginson, Richard. *Called to Account.* Guildford: Eagle, 1993.
Hillary, Ruth. *The Eco-Management and Audit Scheme: A Practical Guide.* Letchworth: Technical Communications (Publishing) Ltd., Undated.
Hines, Tony. *Foundation Accounting.* London: Checkmate/ Arnold, 1987.
Hinton, Teresa and Berthoud, Richard. *Money Advice Services.* London: Policy Studies Institute, 1988.
Hoffman, Michael, Frederick, Robert and Petry, Edward S. Jr. (eds.) *The Corporation, Ethics, and the Environment.* Westport: Quorum Books, 1990.
Hoffman, W. Michael, Kamm, Judith Brown, Frederick, Robert E. and Petry, Edward S. Jr. (eds.) *Emerging Global Business Ethics.* London: Quorum Books, 1994.
Howells, Geraint, Crow, Iain, and Moroney, Maureen. *Aspects of Credit and Debt.* London: Sweet & Maxwell, 1993.
Hughes, M.A. *The Changing Lending Role of Managers in the Financial Services Sector.* Brighton Polytechnic: M.Phil., 1990.
Humble, John. *Social Responsibility Audit: A ManagementTool for Survival.* London: Foundation for Business Responsibilities, 1973.
Hutton, Will. *The State We're In.* London: Jonathan Cape, 1995.
Hwang, B-H. *The Theology of Property: a Consideration of Christian Responsibility of Ownership and Use in the Bible and Tradition.* Cardiff: Ph.D., 1995.
Institute of Business Ethics. *Report on Business Ethics Codes, 1996.* London: IBE, 1996.
Jack, R.B. (Chairman). *Banking Services: Law and Practice. Report by the Review*

Committee. London: HMSO, 1989.

Jennings, Marianne M. *Case Studies in Business Ethics.* Minneapolis/St. Paul: West Publishing Company, 1993.

Johnson, Harold L. 'Can the Businessman Apply Christianity?' *Harvard Business Review*, September-October, 1957 reprinted in *Ethics for Executives.* USA: Harvard Business School, 1968.

Jones, Russell A. *Research Methods in the Social and Behavioral Sciences.* Sunderland, Massachusetts: Sinauer Associates, 1985.

Kant, Immanuel. *Critique of Practical Reason: And Other Writings in Moral Philosophy.* Translated by Beck, Lewis W. Chicago: The University of Chicago Press, 1949.

Kempson, Elaine and Rowlingson, Karen. *Gas Debt and Disconnection.* London: Policy Studies Institute, 1993.

Kempson, Elaine. *Money Advice and Debt Counselling.* London: Policy Studies Institute, 1995.

Keynes, John M. *The General Theory of Employment, Interest and Money.* London: Macmillan, 1964.

Koford, Kenneth and Penno, Mark. 'Accounting, Principal-Agent Theory, and Self-Interested Behavior'. In (eds.) Bowie, Norman E. and Freeman, R. Edward, *Ethics and Agency Theory: An Introduction.* Oxford: Oxford University Press, 1992.

Kristol, Irving. 'A Capitalist Conception of Justice'. In (eds.) Hoffman, W. Michael and Moore, Jennifer Mills *Business Ethics: Readings and Cases in Corporate Morality.* New York: McGraw-Hill, 1990.

Kuhn, James W. and Shriver, Donald W., Jr. *Beyond Success: Corporations and their Critics in the 1990s.* New York and Oxford: Oxford University Press, 1991.

Lane, Robert E. *The Market Experience.* Cambridge: Cambridge University Press, 1991.

Lang, Raymond. 'Analysis of Individuals with Multiple Debt Problems.' In (ed.) Hartropp, Andrew, *Families in Debt.* Cambridge: Jubilee Centre Publications Ltd., 1988.

Lehman, Cheryl R. *Accounting's Changing Role in Social Conflict.* London: Paul Chapman Publishing Ltd., 1992.

Levitt, Theodore. 'The Dangers of Social Responsibility.' *Harvard Business Review*, September-October, 1958. Reprinted in (eds.) Beauchamp, Tom L. and Bowie, Norman E. *Ethical Theory and Business.* Englewood Cliffs, New Jersey: Prentice-Hall, Inc. 1979.

Lloyd, Lord, of Hampstead and Freeman, M.D.A.(eds.). *Lloyd's Introduction to Jurisprudence.* London: Stevens & Sons, 1985.

Lomax, David F. 'Policy for Saving in the 1990s.' pp.14-28. *National Westminster Bank Quarterly Review,* August, 1990.

'Lord Ezra's Conclusions', p.7, *CSA News,* Summer, 1990.

Lucey, T. *Management Accounting.* London: DP Publications Ltd, 1992.

Lynch, James J. *Ethical Banking: Surviving in an Age of Default.* London: MacMillan, 1991.

MacIntyre, Alasdair. *After Virtue: a Study in Moral Theory.* London: Duckworth, 1981.
Maclennan, Duncan, Gibb, Kenneth and Stephens, Mark. 'The Future Sustainability of Home-ownership', pp.1-4, *Findings: Housing Research 228,* December, 1997.
Madura, Jeff. *Financial Markets and Institutions.* Third Edition. Minneapolis: West Publishing Company, 1995.
Madura, Jeff. *International Financial Management.* St. Paul: West Publishing Company, 1995.
Malone, Thomas F. 'Stewardship: Paradigm for the Nth Millennium.' pp.2-3. *National Forum,* vol.75, no.1.
Marshall, Paul. 'Towards and Christian View of State and Economy.' In (ed.) Chaplin, Jonathan. *When Christians Disagree: Politics and the Parties.* Leicester: Inter-Varsity Press, 1992.
Martins, H.B. *Risk Management of U.S. Banks in Less Developed Countries: A Country-risk Analysis.* Sheffield: Ph.D., 1991.
Mattessich, Richard. 'Paradigms, Research Traditions and Theory Nets of Accounting.' In (eds.) Mumford, M.J. and Peasnell, K.V. *Philosophical Perspectives on Accounting: Essays in Honour of Edward Stamp.* London: Routledge, 1993.
Matthews, John B., Goodpaster, Kenneth E., and Nash, Laura L. *Policies and Persons: A Casebook in Business Ethics.* New York: McGraw-Hill, Inc., 1991.
McQueen, John. *What to do when Someone has Debt Problems: A Practical Survival Guide.* Kingswood, Surrey: Elliot Right Way Books, 1985.
Miller, Morris. *Debt and the Environment: Converging Crisis.* New York: United Nations Publications, 1991.
Millichamp, A.H. *Foundation Accounting.* London: DP Publications Ltd., 1989.
Mills, Paul. *The Ban on Interest: Dead Letter or Radical Solution?* Cambridge Papers, vol.2, no.1, 1993.
Mills, P. *Should Interest Exist? - Non-usurious Finance in Economic Thought, Theory and Practice.* Cambridge: Ph.D., 1995.
Mills, Paul. *Interest in Interest: The Old Testament Ban on Interest and its Implications for Today.* Cambridge: Jubilee Centre Publications Ltd., Undated.
Mohon, W.R. *The Dignity of Man and Business Enterprise in Reformed Thinking.* Durham: M.A., 1987.
'Money Advice Initiative.' p.11. *Credit Management,* May, 1989.
Moyer, R. Charles, McGuigan, James R. and Kretlow, William J. *Contemporary Financial Management.* Minneapolis: West Publishing Co., 1995.
Mumford, M.J. and Peasnell, K.V. (eds.) *Philosophical perspectives on accounting: Essays in honour of Edward Stamp.* London and New York: Routledge,
Murphy, Patrick E. 'Corporate Ethics Statements: Current Status and Future Prospects.' pp. 727-740. *Journal of Business Ethics,* 14, 1995.
Naser, Kamal H.M. *Creative Financial Accounting: Its Nature and Use.* London: Prentice Hall, 1993.
National Consumer Council. *Credit and Debt: The Consumer Interest.* London: HMSO, 1990.
National Consumer Council. *Debt Advice Provision in the United Kingdom: A Report*

by Tim Young of the Community Information Project. London: National Consumer Council, 1990.

National Consumer Council. *Appendices to Debt Advice Provision in the United Kingdom.* London: National Consumer Council, 1990.

National Consumer Council. *NCC Annual Report.* London: National Consumer Council, 1990.

National Consumer Council. *Security Risks: Personal Loans Secured on Homes.* London: National Consumer Council, 1987.

Noonan, John P. *General & Special Ethics.* Chicago: Loyola University Press, 1963.

North, Gary. *An Introduction to Christian Economics.* USA: The Craig Press, 1974.

Novak, Michael. *The Spirit of Democratic Capitalism.* London: The IEA Health and Welfare Unit, 1991.

Nozick, Robert. 'The Entitlement Theory'. In (eds.) Beauchamp, Tom L. and Bowie, Norman E. *Ethical Theory and Business,* Fourth Edition. Englewood Cliffs: Prentice-Hall Inc., 1993.

Office of Fair Trading. *CreditWise: Your guide to trouble-free credit.* London: HMSO, Undated.

Office of Fair Trading. *Debt: A Survival Guide.* London: HMSO, Undated.

Office of Fair Trading. *Moneyfax: The Crucial Guide to Credit and Debt.* London: HMSO, 1989.

Office of Fair Trading. *Overindebtedness: A Report by the Director General of Fair Trading.* London: OFT, 1989.

Office of the Banking Ombudsman, The. *The Banking Ombudsman Scheme Annual Report, 1996-1997.* London: The Office of the Banking Ombudsman, 1997.

Outka, Gene and Reeder, John P. Jr. (eds.) *Prospects for a Common Morality.* Princeton, New Jersey: Princeton University Press, 1993.

Page, M. J., *Aspects of Financial Reporting Practices by Companies and of the Accountancy Profession in the United Kingdom.* Southampton: Ph.D., 1990.

Park, S.Y. *Reinvestigations of Risks of Sovereign Loans to Developing Countries.* Liverpool: Ph.D., 1993.

Parker, David. 'Chaos Theory and the Management of Change.' Unpublished Paper dated May 1994.

Passmore, John. *Man's Responsibility for Nature: Ecological Problems and Western Traditions.* London: Duckworth, 1980.

Patel, Natu, (ed.). *Central Statistical Office: Financial Statistics.* No. 406. London: HMSO, 1996.

Peacocke, Arthur R. *Creation and the World of Science: The Bampton Lectures, 1978.* Oxford: Clarendon Press, 1979.

Pizzey, Alan. *Accounting and Finance: a Firm Foundation.* London: Cassell, 1986.

Pointon, F.W. 'Conflicts of Interest for Lending Bankers'. In (eds.) Prindl, Andreas R. and Prodhan, Bimal. *Ethical Conflicts in Finance,* Oxford: Blackwell, 1994.

Preston, Ronald H. *The Future of Christian Ethics.* London: SCM Press Ltd, 1987.

Preston, Ronald H. *Religion and the Ambiguities of Capitalism.* London: SCM Press, 1991.

Prindl, Andreas R. and Prodhan, Bimal (eds.). *Ethical Conflicts in Finance.* Oxford: Blackwell, 1994.

Prodhan, Bimal. 'Ethics, Finance and Society'. In (eds.) Prindl, Andreas R. and Prodhan, Bimal. *Ethical Conflicts in Finance,* Oxford: Blackwell, 1994.

Ragkhan, O.A.W. *The International Debt Crisis and the Economics of Islam.* Aberystwyth: MPhil., 1991.

Raiborn, Cecily A., Barfield, Jesse T. and Kinney, Michael R. *Managerial Accounting,* St. Paul, Minneapolis: West Publishing Company, 1993.

Rawls, John. *A Theory of Justice.* Oxford: Oxford University Press, 1992.

Rawls, John. 'Justice as Fairness'. In (eds.) Hoffman, W. Michael and Moore, Jennifer Mills *Business Ethics: Readings and Cases in Corporate Morality.* New York: McGraw-Hill, 1990.

Riahi-Belkaoui, Ahmed. *Morality in Accounting.* London: Quorum Books, 1992.

Richardson, Bill and Curwen, Peter. 'Do Free-Market Governments Create Crisis-Ridden Societies?' pp.551-560. *Journal of Business Ethics* vol. 14, 1995.

Richter, Alan and Barnum, Cynthia. 'When Values Clash'. pp. 42-45. *HRMagazine,* September, 1994.

Rickwood, Colin and Thomas, Andrew. *An Introduction to Financial Accounting.* London: McGraw-Hill Book Company, 1992.

Rossouw, G.J. 'Rational Interaction for Moral Sensitivity: A Postmodern Approach to Moral Decision-making in Business.' pp.11-20. *Journal of Business Ethics* 13, 1994.

Rouse, C. Nicholas. *Bankers' Lending Techniques.* London: The Chartered Institute of Bankers, 1990.

Rowlingson, Karen and Kempson, Elaine. *Paying With Plastic: A Study of Credit Card Debt.* London: Policy Studies Institute, 1994.

Sales, Stephen M. 'Threat as a Factor in Authoritarianism: An Analysis of Archival Data.' pp.44-57. *Journal of Personality and Social Psychology,* vol. 28. (1973)

Samli, A. Coskun. *Social Responsibility in Marketing: A Proactive and Profitable Marketing Management Strategy.* London: Quorum Books, 1992.

Samuels, J. M. and Wilkes, F. M. *Management of Company Finance.* Wokingham: Van Nostrand Reinhold (UK) Co. Ltd., 1986.

Schaeffer, Francis A. *Pollution and the Death of Man: The Christian View of Ecology.* London: Hodder and Stoughton, 1970.

Schofield, John A. *Cost-Benefit Analysis In Urban & Regional Planning.* London: Allen & Unwin, 1987.

Seaver, Paul. 'The Puritan Work Ethic Revisited.' pp.35 - 53. *Journal of British Studies,* 19(2), 1980.

Selby, Peter. *Grace and Mortgage: The Language of Faith and the Debt of the World.* London: Darton, Longman and Todd Ltd., 1997.

Selden, Raman. 'The Rhetoric of Enterprise'. In Keat, Russell and Abercrombie, Nicholas. *Enterprise Culture.* London: Routledge, 1991.

Sen, Amartya. *On Ethics and Economics.* Oxford: Basil Blackwell, 1988.

Shapiro, Alan C. *Foundations of Multinational Financial Management.* Boston: Allyn and Bacon, 1994.

Sheard, Martin. 'New problems for lenders.' p.21. *Banking World*, September,1992.

Shurman, Laurence. 'What is Fair in the Relationship Between a Bank and its Customer?' *The Banks and Society,* London: The Chartered Institute of Bankers, 1991.

Slaughter, Richard A. *The Foresight Principle: Cultural Recovery in the 21st Century.* London: Adamantine Press Limited, 1995.

Sleeman, John F. 'Usury and Interest.' In Childress, James F. and MacQuarrie, John (eds.) *A New Dictionary of Christian Ethics.* London: SCM, 1986.

J.J.C. Smart and Bernard Williams. *Utilitarianism For and Against.* Cambridge: Cambridge University Press, 1973.

Smart, J. J. C. 'Distributive Justice and Utilitarianism.' In Hoffman, W. Michael and Moore, Jennifer Mills *Business Ethics: Readings and Cases in Corporate Morality.* New York: McGraw-Hill, 1990.

Solomon, Robert C. *Ethics and Excellence: Cooperation and Integrity in Business.* New York and Oxford: Oxford University Press, 1992.

Sorell, Tom and Hendry, John. *Business Ethics.* Oxford: Butterworth-Heinemann Ltd, 1994.

Stoeckle, Bernhard (ed.). *Concise Dictionary of Christian Ethics.* London: Burns & Oates, 1979.

Stone, Christopher D. 'Should Trees have standing? - Toward Legal Rights for Natural Objects.' *Southern California Law Review*, vol. 45(1972) pp. 453-460, 463-464, 480-481,486-487. In (eds.) Beauchamp, Tom L. and Bowie, Norman E. *Ethical Theory and Business*. Englewood Cliffs: Prentice-Hall, 1979.

Sunderland, K. 'Sovereign Debt- Sweet Dreams or Nightmares?' p.4. *Banking World,* December, 1988.

Tawney, R.H. *Religion and the Rise of Capitalism: A Historical Study.* London: John Murray, 1964.

Tondeur, Keith. *Escape from Debt.* Cambridge: Credit Action, 1993.

Tondeur, Keith. *Say Goodbye to Debt.* London: Marshall Pickering, 1994.

Vallely Paul, *Bad Samaritans, First World Ethics and Third World Debt.* London: Hodder and Stoughton, 1990.

Van Til, Cornelius. *The Defense of the Faith.* Philadelphia, Pennsylvania: The Presbyterian and Reformed Publishing Company, 1976.

Van Til, Cornelius. *In Defense of the Faith, Volume III: Christian Theistic Ethics.* Philadelphia: Den Dulk Christian Foundation, 1971.

Vickers, Douglas. *Economics and Man: Prelude to a Christian Critique.* USA: Craig Press, 1976.

Walton, Clarence C. (ed.) *The Ethics of Corporate Conduct.* Englewood Cliffs, New Jersey: Prentice-Hall, Inc., 1977.

Walzer Michael. *Spheres of Justice: A Defense of Pluralism and Equality.* USA: Basic Books, Inc., 1983.

Watson, George. *The English Ideology: Studies in the Language of Victorian Politics.* London: Allen Lane, 1973.

Webb, Eugene J., Campbell, Donald T., Schwartz, Richard D. and Sechrest, Lee. *Unobtrusive Measures: Nonreactive Research in the Social Sciences.* Chicago: Rand McNally & Company, 1971.

Weber, Max. *The Protestant Ethic and The Spirit of Capitalism.* London: Routledge, 1992.

Werhane, Patricia. *Adam Smith and His Legacy for Modern Capitalism.* Oxford: Oxford University Press, 1991.

Weston, J. Fred and Copeland, Thomas E. *Managerial Finance.* Adapted from the eighth American edition by (eds.) Limmack, Robin J. and Fox, Alan F. London: Cassell, 1988.

White, Lynn, Jr. 'The Historical Roots of our Ecologic Crisis'. pp.1203-1207. *Science,* vol. 155, March 1967.

Whyley, C., Kempson, E. and Herbert, A. *Money Matters: Approaches to Money Management and Bill Paying.* London: Policy Studies Institute, 1997.

Williams, Gerald J. *Ethics in Modern Management.* New York: Quorum Books, 1992.

Williamson, Duncan. *Cost and Management Accounting,* London: Prentice Hall, 1996.

Wolfe, Mike. *Debt Advice Handbook.* London: CPAG Ltd., 1995.

Woods, Ngaire. *Ethics and Interests in the International Political Economy, the Management of the Mexican Debt, 1982 - 1989.* Oxford: DPhil., 1992.

Ability, 166, 179
Accountability, 4, 56, 79-80, 95, 114-116, 139, 144, 152, 157, 183-184, 200-203, 207
Accountancy, 2, 52-57
Adams, R.M., 96
Adams, T., 122
Advance application form, 135-136
Advance control, 30
Advertising, 112
Agency Theory, 55-57, 64, 76-77, 78
Altruism, 58
Andrews, D, 116
Andrews, K.R., 61, 65
APR, 176
Aristotle, 12
Asset enhancement, 100, 167-168
Assets, 107
Atherton, J., 25
Audit, 107

Balance Sheet, 53-54, 135
Banker-customer relationship, 130, 139
Banking Education Service, 65
Banking ethics, 6
Banking World, 141, 146, 149-150
Bankruptcy, 14, 15
Bankruptcy Association, 35
Banks, 129
Barcena, A., 154
Barlow Clowes, 60
Barty-King, H., 14
Base rate, 29
BCCI, 60
Beauchamp, T.L., 62, 63, 161, 162
Berthoud, Richard 35, 116, 118, 119, 177-178
Beutler, I.F., 36
Block, P., 1, 69, 85, 87, 94, 99
Blue Arrow, 60
Boleat, M., 112, 123
Bolton Report, 133-134
Borrie, Sir Gordon, 15, 42
Borrower-lender relationship, 39-42
Bottom-line, 38
Bowie, N.E., 62, 63, 161, 162
Brady, N., 31
Briggs, A., 23
Bromwich, M., 54
Brown, H.S., 38
Browne, A.H., 149-150
Budget deficits, 31-32
Bush, G., 68
Business enterprise, 83-88
Business Ethics, 7
Business Ethics literature, 5-6
Business finance, 131-141

Cadbury, Sir A., 26
Calling, 18
CAMPARI, 131-132, 141-142, stewardship evaluation, 132-133
Cannon, T., 1, 3, 109
Canon Law, 14
Canons of lending, 131-133
Capitalism, 16-19, 43
Carr, S. 153
Cash flow statement, 53, 135
Causation, 121
Cause and effect, 22
Certo, S.C., 26
Chaos Theory, 22
Checkley, K., 135
Chemical Manufacturers Association, 70
Christian values, 4
Citizen's Advice Bureaux, 124, 175, 178
Codes, 7, 8, 24, 27
Collins, N., 174
Common, M., 81
Common morality, 66-69, 96
Companies, 3
Competitive market, 17
Congdon, T., 28, 29, 31, 32, 33, 173-174
Conservation, 74, 81, 153
Constituencies, 24,26,38,189
Consultation, 26
Consumer credit, 16, 121

Consumer Credit Counselling Service, 183, 184, 207-208
Consumer lending, 141-158
Core values, 25, 26, 66, 74-80
Corporate governance, 3, 69, 94
Corporate philanthropy, 25
Corporate purpose, 24-25
Cost Benefit Analysis, 81-82, 99
Cowton, C.J., 27
Credit, 15, 16, 118-120
Credit Action, 177
Credit creation, 23
Credit limit, 119
Credit procedures, 171
Credit promotion, 34, 93, 124-125, 152
Credit references, 145-147
Credit scoring, 7, 41, 145
Crowther Committee, 119

Daft, R.L., 3
Damnum energens, 13
Davies, A., 173-174
Debt, 119
Debt forgiveness, 31, 38, 179-181
Debt/income ratio, 28
Debt management model, 191-208
Debt management plans, 183
Debt remission, 30-31
Debt repudiation, 40, 42
Debtors Act, 1730, 14
Decision support model, 1, 7
Dees, J.G., 157-158, 174, 177
Defoe, D., 14
Depositor protection, 202-203
Depositors, 129
Deregulation, 123
Development, 79, 95, 109-114, 138-139, 143, 152, 157, 167, 198-200, 207
Distributive justice, 161-166
Dobson, A., 109-110
Dodds, W., 15
Domestic budgeting, 115-116, 152
Donaldson, J., 7, 66, 67, 117
Donaldson, T., 99, 182-183
Doyle, E.P., 130
Earnings per share, 54
Ecology, 73
Economic activity model, 164-165
Economic dynamics, 13
Economic Man, 19
Economic modelling, 21
Economics, 16-19, 189
Effective demand, 32
Effectiveness, 25, 57, 79-80, 95, 106-109, 137-138, 143, 151, 156, 170-171, 181-182, 194-195, 206
Effort, 163, 166, 179
Egoism, 62, 64
Elliott, B., 54
Elster, J., 21
Emotional crisis, 92-93
Empirical data, 4, 5
Empirical research, 34
Enhancement, 80
Entrepreneurial activity, 17
Entrustedness, 2, 4, 56, 74, 79-80, 85, 95, 96-98, 129, 137, 143, 150-151, 154-155, 193-194, 205
Environmental protection, 3, 68
Environmental stewardship, 72-75, 77, 80-82
Environmental strategy, 3
Equality, 162, 163
Equity withdrawal, 173-174
Ethical analysis, 56
Ethical canons, 6
Ethical debt management, 93-95
Ethical debt management literature, 5
Ethical investment, 25, 67
Ethical model, 43
Evan, W.M., 26, 154
Expenditure, 92
Experimentation, 4

Fairholm, G.W., 106
Fairness, 79-80, 95, 117, 139-140, 144, 152-153, 157, 177-178, 179, 204
Faith in the City, 105

Family stewardship, 102-106
Family welfare, 103, 111
Fayol, H., 4
Fictitious funds, 13
Final accounts, 3
Financial accounting, 2, 52
Financial growth, 84
Financial information, 53, 131, 133-134
Financial institutions, 93, 129
Financial interrelations ratio, 23, 34
Financial measurement, 32
Financial system, 14, 43
Fiscal policy, 23
Fleming, M.C., 123
Flexible borrowing, 120
Ford, J., 35, 107, 112, 116, 119, 121, 170, 171
Fractional reserves, 13
Free market doctrine, 24-25
Free Trade Debate, 1846, 23
Freedley, E., 45
Freeman, R.E., 1
Friedman, M., 23, 61, 65, 72
Friends Provident, 69
Futurism, 18-19

Galbraith, K., 169
Gardner, E.C., 161
Garnett, E.J., 45, 46
Garrison, S.H., 21
Geisler, N.L., 74
George, S., 31
Gilbert, D.R., 5, 117
Goals, 168-169
Goldsmith, O., 15
Good will, the, 8
Government borrowing, 39
Government policy, 23, 174-175
Government policy effects, 112
Gray, R., 107
Griffiths, B., 114
Guinness, 59

Hanks, R.S., 36
Hanna, R., 93, 108
Hard-core borrowing, 40
Hartley, R.F., 156
Hartropp, A., 33, 35
Harvard Business Review, 71
Hayek, F.A., 23, 170
Henderson, H., 18
Henry, M., 50-51
Higginson, R., 105
History of debt, 14-15
Home ownership, 104, 112, 123, 162, 167
Howells, G., 35
Hutton, W., 37
Hwang, B.H., 103
Hyper-inflation, 32

Income reduction, 91, 107
Inflation, 23, 24, 29-30, 32, 33, 40, 42
Information, 55
Information packs, 134, 156
Institute of Business Ethics, 27, 35
Interest, 12-14, 29
Interest rates, 39, 40, 41, 43
Internalising values, 7, 93-94
Interviews, 134-135
Irrational behaviour, 28, 30
Islam, 181

Joachimson v Swiss Bank Corporation, 130
Johnson, H.L., 71-72
Joint Stock Companies Act, 53
Jubilee Centre, 16, 29, 91, 107, 120, 166
Judaism, 73
Justice, 17, 161-166

Kant, I., 8
Keynesianism, 23, 32, 39
Koford, K., 157
Kolb, R., 4
Kristol, I., 168
Kuhn, J.W., 26, 67-68

Laissez-faire, 18, 23
Lane, R.E., 22
Learning cycle, 4
Lehman, C.R., 54, 56
Lending criteria, 123
Lending procedures, 133-141
Levitt, T., 61
Literature for ethical debt management, 6
Loan applications, 101
Loan appraisal, 141-145, 151-152
Loan purpose, 172-173
Lomax, D.F., 174
Lucrum cessans, 13
Lynch, J.J., 6

MacIntyre, A., 67
Macroeconomics, 30, 34, 122
Madura, J., 29-30, 37
Magna Carta, 14
Malone, T.F., 68
Management, 36, 38, 84
Management accounting, 2-3, 52
Management research, 5
Managerial stewardship, 52, 53, 54, 59-60
Mannion, R., 182
Market place, 17, 20, 27, 82-83, 168
Market theory, 21
Marshall, P., 96-97
Matching, 53, 79-80, 82, 85-87, 95, 98-103, 137, 143, 151, 155-156, 194, 196-198, 205-206
Mattessich, R., 52, 55, 56
Matthews, J.B., 36
McQueen, J., 35, 100, 101
Medium of exchange, 12
Miller, M., 68, 181
Millichamp, A.H., 2
Mills, P., 13
Money advice, 175-184
Money supply, 24
Monitoring, 136-137, 139, 144, 203
Moral values, 18
Mortgage finance, 166-175
Mortgage tax relief, 123
Mortgages, 34, 116, 122, 142
Multiple debtors, 91

Naser, K.H.M., 53
National Consumer Council, 34, 35, 115, 141, 145, 146, 148, 171, 175, 178, 183-184
National Debt, 14, 23, 31-32
National Remortgage Corporation, 124
Nationalised Industries, 18
Need, 162-163, 167, 170, 179
Neild, James, 15
Net income, 54
Noonan, J.P., 17
Novak, M., 50, 66
Nozick, R., 162

Objectives, 86, 113-114
Objects clause, 83, 202
Office of Fair Trading, 121
Oglethorpe, J., 14
Operational model, 5, 38, 87
Opportunity cost, 13
Overcommitment, 28, 39, 41, 43, 122
Overindebtedness, 34
Overspending, 92

Parker, D., 22
Parker, G., 35, 182
Participation, 27
Passmore, J., 70, 81, 82
Peacocke, A.R., 73
Personal borrowers, 110-111
Pitt, M., 14
Pizzey, A., 2
Pointon, F.W., 30
Policy Studies Institute, 35, 115-116
Positive Economics, 20
Positivism, 81
Poverty, 105-106
Preservationist, 74
Preston, R.H., 13, 24, 32, 44-45, 63, 68
Price mechanism, 17, 20, 21, 23

Primitivism, 81
Private ownership, 17
Private property, 17
Product stewardship, 70, 77
Profit goal, 38
Profit margin, 24
Profit maximization, 25, 61, 66
Profit motivation, 17
Protestant Work Ethic, 18, 19, 44, 68
Prudence, 58
Public debt, 33, 43

Ragkham, O.A.W., 181
Rationality, 20-21, 189
Rawls, J., 168-169
Reductionism, 58
Religious stewardship, 70, 77
Remortgages, 124
Resource allocation, 20
Resources, 2, 4
Rhodes, R., 136
Richardson, B., 24
Rickwood, C., 2
Rights, 161-162, 166
Risk assessment, 148
Risk management, 37, 195
Risk premium, 41
Risk-return trade-off, 37
Rouse, N., 131, 135-136, 140, 141, 143

Samli, A.C., 38
Savings deficit units, 1
Savings surplus units, 1
Scientific method, 4
Seaver, P., 50
Security, 37
Security risks, 34
Selby, P., 69, 104, 106, 180
Sen, A., 20, 21
Shaftesbury, Lord, 15
Shapiro, 31
Shareholder returns, 38
Shareholders, 37, 53, 60
Sharp, A., 116
Short-termism, 18
Shriver, D.W., 26, 67-68
Sleeman, J., 13
Small Business Loan Guarantee Scheme, 136
Smith, A., 17, 62
Social responsibility, 46, 61-66, 130, 201-202
Social utility, 165-166, 167, 176-177
Societas, 13
Solomon, R.C., 158, 179
Solvency, 99-101
Something to Celebrate, 103
Sorell, T., 171
Sound money, 23, 43
Specialization, 24
Speculation, 173-174
Stakeholders, 26, 38, 54, 189
Sterling debt, 39-40
Stewardship, 44-46, 58
  and accountancy, 2, 52-57
  and Christian values, 4
  and corporate governance, 3
  and environmental strategy, 3
  concept, 1, 7, 188
  decision-making, 80-82
  history, 2
  operational model, 193
Stewardship Accounting, 53, 75-76
Stewardship values, 8
  in lending, 137-141
Stone, C.D., 72
Strategies, 117-120
Suboptimization, 64
Supply and demand, 21
Supportiveness, 176-177
Surplus income, 101-102
Sustainable development, 154
Synergy, 27
Systems appraisal methods, 145-149

Tawney, R.H., 18
Taylor, F., 4
Testing, 4, 5, 6, 7

Third Lateran Council, 12
Third World Debt, 6, 31, 38
Trading and Profit and Loss Account, 53-54, 135
True and fair view, 84
Trust, 56, 89
Trusteeship, 74

Uncertainty, 120
Usury, 12-14
Utilitarianism, 62-64, 169

Values, 27, 44, 59, 66, 67, 78-80, 190-191
Van Til, 70, 71
Vicegerency, 70-71, 73, 74, 77
Victorian Christians, 46

Walzer, M., 161, 164
Watson, G., 23
Weber, M., 18
Werhane, P., 17
Weston, J.F., 33
White, L., 73
Williams, G.J., 17
Wolfe, M., 104, 105

Young, T., 175

Issues in Business Ethics

1. G. Enderle, B. Almond and A. Argandoña (eds.): *People in Corporations.* Ethical Responsibilities and Corporate Effectiveness. 1990 ISBN 0-7923-0829-8
2. B. Harvey, H. van Luijk and G. Corbetta (eds.): *Market Morality and Company Size.* 1991 ISBN 0-7923-1342-9
3. J. Mahoney and E. Vallance (eds.): *Business Ethics in a New Europe.* 1992 ISBN 0-7923-1931-1
4. P.M. Minus (ed.): *The Ethics of Business in a Global Economy.* 1993 ISBN 0-7923-9334-1
5. T.W. Dunfee and Y. Nagayasu (eds.): *Business Ethics: Japan and the Global Economy.* 1993 ISBN 0-7923-2427-7
6. S. Prakash Sethi: *Multinational Corporations and the Impact of Public Advocacy on Corporate Strategy.* Nestle and the Infant Formula Controversy. 1993 ISBN 0-7923-9378-3
7. H. von Weltzien Hoivik and A. Føllesdal (eds.): *Ethics and Consultancy: European Perspectives.* 1995 ISBN Hb 0-7923-3377-2; Pb 0-7923-3378-0
8. P. Ulrich and C. Sarasin (eds.): *Facing Public Interest.* The Ethical Challenge to Business Policy and Corporate Communications. 1995 ISBN 0-7923-3633-X; Pb 0-7923-3634-8
9. H. Lange, A. Lhr and H. Steinmann (eds.): *Working Across Cultures.* Ethical Perspectives for Intercultural Management. 1998 ISBN 0-7923-4700-5
10. M. Kaptein: *Ethics Management.* Auditing and Developing the Ethical Content of Organizations. 1998 ISBN 0-7923-5095-2; Pb 0-7923-5096-0
11. R.F. Duska (ed.): *Education, Leadership and Business Ethics.* Essays on the Work of Clarence Walton. 1998 ISBN 0-7923-5279-3
12. R. Mohon: *Stewardship Ethics in Debt Management.* 1999 ISBN 0-7923-5747-7

KLUWER ACADEMIC PUBLISHERS – DORDRECHT / BOSTON / LONDON

GPSR Compliance

The European Union's (EU) General Product Safety Regulation (GPSR) is a set of rules that requires consumer products to be safe and our obligations to ensure this.

If you have any concerns about our products, you can contact us on ProductSafety@springernature.com

In case Publisher is established outside the EU, the EU authorized representative is:

Springer Nature Customer Service Center GmbH
Europaplatz 3
69115 Heidelberg, Germany

Batch number: 10823959

Printed by Printforce, the Netherlands